ARTHUR GODMAN &
RONALD DENNEY

BARNES & NOBLE THESAURUS OF **CHEMISTRY**

Consultant: ROGER NORTON

 BARNES & NOBLE BOOKS
A DIVISION OF HARPER & ROW, PUBLISHERS
New York, Cambridge, Philadelphia, San Francisco
London, Mexico City, São Paulo, Sydney

BLA Publishing Limited and the authors would like to thank Roger
Norton for his help in the preparation of the manuscript for this
book.

© BLA Publishing Limited 1985

Library of Congress Cataloging-in-Publication Data

Godman, Arthur
    Barnes & Noble thesaurus of chemistry.

    1. Subject Headings--Chemistry. I. Denney, Ronald. II. Title.
  III. Barnes and Noble thesaurus of chemistry.
  IV. Title: Thesaurus of chemistry.
Z695.1.C5D46   1986   025.4'954   85–45523
ISBN 0-06-463718-2 (pbk.)

This book was designed and produced by
BLA Publishing Limited, Swan Court,
London Road, East Grinstead, Sussex, England.

A member of the **Ling Kee Group**
LONDON·HONG KONG·TAIPEI·SINGAPORE·NEW YORK

Illustrations by Rosie Vane-Wright, Hayward & Martin and
BLA Publishing Limited

Phototypeset in Britain by BLA Publishing Limited and Composing
  Operations Limited
Color origination by Planway Limited
Printed in Spain by Heraclio Fournier

Contents

How to use this book

This book combines the functions of a dictionary and a thesaurus: it will not only define a word for you, but it will also indicate other words related to the same topic, thus giving the reader easy access to one particular branch of the science. The emphasis of this work is on interconnections.

On pages 3 and 4 the contents pages list a number of broad groupings, sometimes with sub-groups, which may be used where reference to a particular theme is required. If, on the other hand, the reader wishes to refer to one particular word there is, at the back of the book, an alphabetical index in which approximately 2000 words are listed.

Looking up one particular word or phrase

Refer to the alphabetical index at the back of the book, then turn to the appropriate page. At the top of that page you will find the name of the general subject printed in bold type, and the specialised area in lighter type. For example, if you look up **inhibitor**[3], you will find it listed on p.188, at the top of which page is **CHEMICAL REACTIONS**/CATALYSIS. If you were unsure of the meaning of the phrase, you may now not only read its definition, but also place it in context. Immediately after the word or phrase you will see in brackets (parentheses) the abbreviation indicating which part of speech it is: (*n*) indicates a noun, (*v*) a verb and (*adj*) an adjective. Then follows a definition of the word, expressed as far as is possible in language which is in common use. Where a related word is listed nearby, a simple system using arrows has been devised.

- (↑) means that the related term may be found above or on the opposite page.
- (↓) means that the related term may be found below or on the opposite page.

A page reference in brackets is given for any word which is linked to the topic but is to be found elsewhere in the book. You will soon appreciate the advantages of this scheme of cross-referencing. Let us take an example. On p.34 the entry **hydrocarbon** is:

hydrocarbon (*n*) any organic chemical compound formed solely from carbon and hydrogen atoms (p.29) combined together. Hydrocarbons may be classified as aliphatic chain structures (p.33), alicyclic (↓) or aromatic (↓).

To gain a broader understanding, the reader will look at the entry **alicyclic** below on the same page, at the entry **aromatic** below on the same page, and will also refer to the entry **atom** on p.29 and **chain structure** on p.33.

Searching for associated words

As the reader will have observed, the particular organisation of this book greatly facilitates research into related words and ideas, and the extensive number of illustrations and diagrams assists in general comprehension.

Retrieving forgotten or unknown information

It would appear impossible to look up something one has forgotten or does not know, but this book makes it perfectly feasible. All that is required is a knowledge of the general area in which the word is likely to occur and the entries in that area will direct you to the appropriate word. If, for example, one wished to know more about the **Joule-Thomson effect**, but had forgotten the term, it would be sufficient to know it was connected with **vapour**; the reader looking up **vapour** would be referred to **critical temperature** which would then indicate **liquefaction of gases**. This, in turn, refers to the **Joule-Thomson effect**, which is defined and/or further explained by means of a diagram.

Studying or reviewing a subject

Two methods of using this book will be helpful to the reader who wishes to know more about a topic, or who wishes to review knowledge of a topic.

(*i*) For a broader understanding of crystals, for example, you would turn to the section dealing with this area and read through the different entries, following up the references which are given to guide you to related words.

(*ii*) If you have studied one particular branch of the science and you wish to review your knowledge, looking through a section on **electromotive force**, by way of example, might refresh your memory or introduce an element which you had not previously realised was connected.

melting point

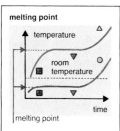

heating curves at
atmospheric pressure

○ liquid
■ solid
▽ liquid + solid
△ molten

freezing point

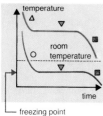

cooling curves at
atmospheric pressure

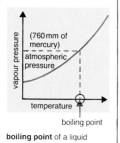

boiling point of a liquid

physical property a property of an object or a radiation. A property of a substance or material for which the chemical nature of the substance or material does not change. Examples of physical properties are: density, state of matter, latent heat, ductility, dielectric constant.

intensive property a property that depends upon the amount of matter, e.g. mass, volume, resistance.

extensive property a property that is independent of the amount of matter, e.g. density, melting point, odour, resistivity, specific heat capacity, etc.

state of matter one of the three states of solid, liquid or gas.

melt (*v*) to change a solid to a liquid.

melt (*n*) an amount of a substance, material, or mixture melted in one process.

molten (*adj*) describes a substance which is liquid because its temperature is high; the substance is solid at room temperature, e.g. iron above 1539°C is molten as it melts at 1539°C; molten iron can be poured into moulds.

freeze (*v*) to change a liquid substance into a solid when the substance is liquid at room temperature, e.g. benzene freezes at 5°C.

solidify (*v*) to change a molten (↑) substance into a solid, usually by cooling.

melting point the temperature at which the solid and liquid phases (p.159) of a substance can exist together at standard atmospheric pressure (101 325 N m^{-2}). Melting point is a property of substances which are solid at room temperature; determination of a melting point can be used to establish the identity of a substance. Abbreviation for melting point is m.p.

freezing point the same temperature as melting point but used for substances that are liquid at room temperature. The term *melting point* can be used for such a substance if it is changing from solid to liquid, e.g. the freezing point of water is 0°C, the melting point of ice is 0°C. The abbreviation for freezing point is f.p.

liquefy (*v*) to change a gas or a solid to a liquid. **liquefaction** (*n*).

condense (*v*) to change a vapour to a liquid by cooling, or by applying pressure, or both. **condensation** (*n*).

boiling point the temperature at which the vapour pressure (p.154) of a liquid is equal to the atmospheric pressure, or other external pressure. At the boiling point evaporation takes place throughout the liquid.

sublime (*v*) (1) to change directly from a solid to a vapour or from a vapour to a solid without forming a liquid. (2) to make a solid substance change to a vapour and back to a solid in order to purify it from other substances which do not sublime. **sublimation** (*n*).

specific heat capacity the number of joules of heat required to raise the temperature of 1 kg of a solid or liquid substance through 1 kelvin. The symbol is *c*, and the units J kg^{-1} K^{-1}. For gases, *see* **specific heat capacity at constant pressure** and **constant volume** (p.155).

molar heat capacity the number of joules of heat required to raise the temperature of 1 mole of a substance through 1 kelvin. The units are J mol^{-1} K^{-1}. Molar heat capacity = c × (molar mass).

atomic heat the number of joules of heat required to raise the temperature of one mole of atoms of an element through 1 kelvin.

latent heat the heat energy given out or taken in when a change of state takes place.

specific latent heat the number of joules of heat taken in or given out when 1 kg of a substance undergoes a change of state without change of temperature. The **specific latent heat of fusion** is measured when a solid substance changes to a liquid or vice versa. The **specific latent heat of vaporization** is measured when a liquid changes to a vapour or vice versa. The symbol is l_f or l_e for fusion or vaporization, and the units are J kg^{-1}

relative vapour density the relative vapour density of a gas, or vapour, is the ratio of the mass of any volume of the gas, or vapour, to the mass of an equal volume of hydrogen, with both volumes measured at the same temperature and pressure. It is a number and has no units.

$$\text{rvd} = \frac{\text{mass of gas}}{\text{mass of same vol. of hydrogen}}$$

$$= \frac{m}{m_H} = \frac{m/V}{m_H/V}$$

(volumes of gas and hydrogen are the same)

$$= \frac{\text{density of gas}}{\text{density of hydrogen}}$$

(same temperature and pressure)

rvd is numerically equal to half the relative molecular mass of a gas or vapour.

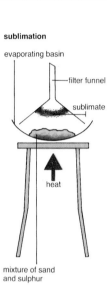

sublimation

evaporating basin

filter funnel

sublimate

heat

mixture of sand and sulphur

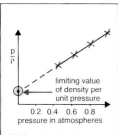

limiting density of a gas

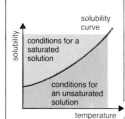

solubility of a crystalline solid

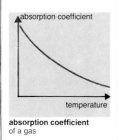

absorption coefficient
of a gas

limiting density the density of a gas is m/V and the density per unit pressure is m/pV, i.e. ρ/p (where ρ is the density). For a given mass of gas, ρ/p is a constant for an ideal gas (p.147). For a real gas the ratio is not constant, but varies with pressure. At pressures below 1 atmosphere, the graph of ρ/p against pressure is close to a straight line for all gases. At zero pressure, all gases become ideal gases in their behaviour. The ratio of ρ/p is extrapolated to zero pressure. The value of ρ/p gives the limiting density value for substitution in the gas equation of state (p.148) from which the relative molecular mass of a gas can be determined.

atomic volume the atomic volume of an element is its relative atomic mass divided by its density. It gives an approximate measure of the size of an atom of an element.

solubility (n) (1) the mass of a solid substance, in grams, that can be dissolved in 100g of a solvent to form a saturated solution in the presence of excess solute at a given temperature. (2) the number of moles of a solid or liquid substance dissolved in 100 moles of solvent in the presence of excess solute at a given temperature. (3) the volume of gas which will just saturate a unit volume of liquid, the gas volume being measured at the temperature and pressure at which the solubility is determined. The solubility varies with pressure and temperature.

absorption coefficient the absorption coefficient of a gas is the volume of gas which will just saturate a unit volume of liquid, the gas volume being measured at s.t.p. The absorption coefficient is independent of pressure, but varies with temperature.

crystalline (*adj*) describes a solid which has a long-range regular lattice structure. *See* **crystal lattice** (p.165).

amorphous (*adj*) describes a solid which has no crystalline structure. Such solids are polymers or glasses (↓).

polymer (n) an amorphous solid consisting of long organic molecules joined together by weak bonds. It has no regular long-range structure.

glass (n) an amorphous solid with individual atoms or molecules with a short-range regular arrangement. The structure of glasses is similar to that of liquids, but their ability to flow is negligible. Soda glass and potash glass are examples of this type of solid.

state of division a description of the size of particles into
which a solid has been divided. The descending order
of size is: lump, chip, flake, granule, grain, powder.
Metals are described as massive, turnings or filings,
e.g. massive zinc, copper turnings, iron filings. Grains,
powders and filings can be coarse or fine; fine
describes the smallest state of division.

elastic (*adj*) describes a solid which returns to its
original shape after a distorting force has stopped acting,
e.g. rubber is elastic. *See* **malleable** (↓). **elasticity** (*n*).

plastic (*adj*) describes a solid which is permanently
deformed by a stress and does not return to its original
shape, e.g. wax is plastic. **plasticity** (*n*).

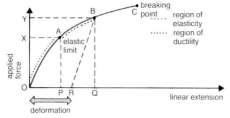

applied force ≤ X
distortion occurs
returns to original length
when force removed

applied force > X
deformation occurs

force Y produces extension OQ
when force removed extension
returns to R or is the deformation

BR is parallel to OA

brittle (*adj*) describes a solid which breaks into pieces
when a stress is applied, e.g. glass is brittle.
brittleness (*n*).

ductile (*adj*) describes a solid material which can be
drawn out by a force to form a thin wire. *See* **malleable**
(↓). **ductility** (*n*).

malleable (*adj*) describes a solid material which can be
beaten flat to form a thin sheet. Elastic solids become
ductile and malleable when sufficient stress is applied
so that the elastic limit of a material is exceeded. Above
the elastic limit the solid becomes plastic. The elastic
limits for ductility and malleability differ, but are of
comparable values. **malleability** (*n*).

lustre (*n*) a quality of a surface which is shining by
reflected light. **lustrous** (*adj*).

hardness (*n*) a quality of a solid. It is a measure of the
elastic properties of a material; the indentation hardness is measured for a simple determination of elasticity.

Mohs' scale a scale of hardness of materials. The
material is tested with each of ten test substances to
see which scratches the surface. The material is then
placed on a scale of hardness of 1 to 10. Scale 10
marks the hardest material.

Mohs' scale of hardness

scale	test material
1	talc
2	rock salt
3	calcite
4	fluorite
5	apatite
6	feldspar
7	quartz
8	topaz
9	corundum
10	diamond

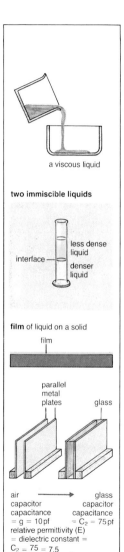

a viscous liquid

two immiscible liquids

less dense liquid

interface

denser liquid

film of liquid on a solid

film

parallel metal plates

glass

air capacitor capacitance = g = 10 pf

glass capacitor capacitance = C_2 = 75 pf

relative permittivity (E) = dielectric constant = $\dfrac{C_2}{C_1} = \dfrac{75}{100} = 7.5$

refractory (*adj*) describes materials that are resistant to heat at high temperatures, and to corrosion by chemical action from substances in contact with them.

mobile (*adj*) describes a liquid that flows readily, e.g. ethanol is a mobile liquid as it moves easily when its container is tilted. **mobility** (*n*).

viscous (*adj*) describes a liquid that does not flow readily, e.g. hydrocarbon oils tend to stick to the side of a container when it is tilted.

viscosity (*n*) the resistance to motion offered by a liquid. The liquid resists flow, and it also resists the motion of a solid body through it. Viscosity is caused by cohesive forces between molecules of the liquid. The symbol for viscosity is η and it is measured in $N s m^{-2}$.

miscible (*adj*) describes liquids which can be mixed to form a single liquid. The result can be considered to be a solution of one liquid in another liquid, e.g. ethanol is miscible with water.

immiscible (*adj*) describes liquids which do not mix. Two immiscible liquids form layers with an interface between them. The less dense liquid is above the denser liquid, e.g. a layer of water above a layer of tetrachloromethane.

interface (*n*) the boundary between phases (p.159) in a heterogenous (p.161) system, e.g. the boundary between two immiscible (↑) liquids; the surface of a liquid when in contact with its vapour.

film (*n*) a thin layer. A liquid film can form on a solid or on another liquid. A gaseous film can form on a solid.

volatile (*adj*) describes a liquid that readily evaporates.

magnetic (*adj*) of, or to do with, magnets; describes a field of force that has the power to attract certain materials; describes materials that are affected by magnets. *See* **ferromagnetism** (p.94).

dielectric constant a dielectric placed between the plates of a capacitor increases its capacitance. The dielectric constant is the ratio of the capacitance of the capacitor with the dielectric between its plates to the capacitance of the capacitor with a vacuum between the plates. An air capacitor gives an approximate result when substituted for a vacuum capacitor (error is 0.05%). Dielectric constant for soda glass is 7.5, e.g. an air capacitor has a capacitance of 10 microfarads, with glass between the plates it is 75 microfarads.

relative permittivity the term now used instead of dielectric constant (↑). Its symbol is ε_r.

chemical property those properties which describe the effect of other substances in changing the named substance, and the effect of heat, light, and electric current when producing a chemical change.

nature (*n*) all the essential properties of a substance or material that make it that particular substance or material. The nature of a form of energy is also described by its essential properties, e.g. the exhibition of allotropy is part of the nature of carbon; the ability of electromagnetic waves to suffer refraction is part of their nature.

labile (*adj*) describes any substance or material that readily undergoes a chemical or physical change. A labile compound decomposes easily.

inert (*adj*) a substance which has no chemical reaction with other substances. **inertness** (*n*).

passive (*adj*) to be unreactive because a thin layer of oxide formed on the surface of a metal is preventing chemical action. **passivity** (*n*).

affinity (*n*) chemical attraction for another substance leading to a strong, or even violent, reaction, e.g. chlorine has an affinity for phosphorus, it reacts strongly; hydrogen chloride gas has a strong affinity for water.

corrosion (*n*) a chemical process in which the surface of a metal is destroyed by the action of certain substances. The metal surface is covered with small holes. Rusting is a form of corrosion.

corrosive (*adj*) describes any substance that readily attacks and destroys the surface of living things and non-living materials. A corrosive substance does not necessarily cause corrosion.

caustic (*adj*) describes any substance that attacks and destroys living tissue, e.g. skin. **causticity** (*n*).

mild (*adj*) describes a chemical agent that is not as powerful as a strong agent, but is more powerful than a weak one, e.g. a strong oxidizing agent oxidizes ethanol to ethanoic acid; a weak oxidizing agent has no effect.

bland (*adj*) describes any substance or material that is soothing, i.e. the opposite of caustic (↑).

noble (*adj*) describes the gases helium, neon, argon, krypton, xenon and radon. They exhibit very weak or no chemical reactivity and have very low boiling points.

permanent (*adj*) (1) describes gases that are not easily liquefied. Such gases have critical temperatures (p.150) much lower than normal atmospheric temperatures. (2) describes water that cannot be softened by boiling.

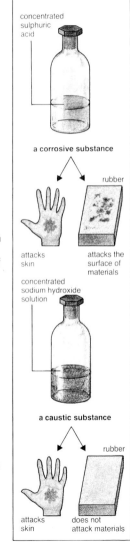

concentrated sulphuric acid

a corrosive substance

attacks skin

attacks the surface of materials

rubber

concentrated sodium hydroxide solution

a caustic substance

attacks skin

does not attack materials

rubber

macrostructure (*n*) the external and observable forms of structure, e.g. the cubic crystal form of sodium chloride is one feature of its macrostructure.

polymorphism (*n*) the phenomenon of solid elements and compounds having more than one crystalline form, e.g. ammonium chloride has two crystalline forms, one with caesium chloride structure, and one with a sodium chloride structure. There are two kinds of polymorphism, enantiotropy (↓) and monotropy (p.14).

allotropy (*n*) the phenomenon of an element having two or more forms without a change of state between the forms. If the forms are crystalline, then they are polymorphs as well as allotropes, e.g. phosphorus has two crystalline forms. If the forms are gaseous or liquid, then they are allotropes and not polymorphs, e.g. oxygen exists as O_2 and O_3. **allotropic** (*adj*).

dynamic allotropy allotropy (↑) in which the allotropes are in dynamic equilibrium, i.e. continually changing from one to another, but maintaining the same equilibrium mixture under the same conditions, particularly temperature. In dynamic allotropy, the difference between allotropes is in the number of atoms in a molecule, e.g. $S_8 \rightleftharpoons S_4$ in liquid sulphur.

allotrope (*n*) one form of an element which exhibits allotropy. Allotropes differ in physical properties, e.g. density, crystalline structure, colour, and may differ in chemical properties, e.g. stability, affinity for oxygen.

enantiotropy (*n*) a type of polymorphism (↑) with a reversible change between the crystalline forms. Each form is stable over a particular temperature range only. The change from one form to another takes place at a transition temperature (↓), e.g. rhombic sulphur changes to monoclinic sulphur at 95.5°C, the change is reversible; red HgI_2 changes reversibly to the yellow form at 126°C. Enantiotropic changes take place slowly; rapid heating or cooling may cause a direct change from one crystalline form to a liquid without passing through the intermediate crystalline form. **enantiotropic** (*adj*).

transition temperature the temperature at which one enantiotropic (↑) polymorph changes into another. An element or compound can have more than one transition temperature, e.g.

Grey tin $\underset{}{\overset{13°C}{\rightleftharpoons}}$ White tin $\underset{}{\overset{161°C}{\rightleftharpoons}}$ Rhombic tin Changes at a transition temperature are slow, so a polymorph can exist in a metastable (p.15) condition outside its temperature range.

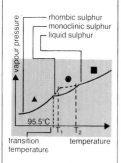

rhombic sulphur
monoclinic sulphur
liquid sulphur

vapour pressure

95.5°C

T_1 T_2

transition temperature

temperature

vapour pressure/temperature curves for the allotropes of sulphur (typical for an enantiotropic system)

▲ for slow heating: m.p. is T_2
● for rapid heating: m.p. is T_1
■ slow and rapid cooling follow the same curves

monotropy (*n*) a type of polymorphism (p.13) in which no reversible change can take place between two crystalline forms. One form is stable and one form is metastable (↓). The metastable form will always change to the stable form at any temperature, although the change will be very slow. A hypothetical transition point would be above the melting point of both forms. Rapid cooling of the liquid substance produces the metastable crystalline form, while slow cooling produces the stable form. The stable form cannot be converted directly to the metastable form; it must first pass through the liquid phase (p.159). Examples of monotropy are: yellow phosphorus (metastable), red phosphorus (stable); graphite (stable), diamond (metastable). The metastable form has the higher energy content in all cases of monotropy. **monotropic** (*adj*).

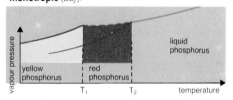

isomorphism (*n*) the phenomenon of different substances forming crystals in which geometrically similar structural units are arranged in similar ways. Crystals of different substances which exhibit the same external structure are commonly said to be isomorphous, yet their internal structure may not be identical. Alums exhibit the same external crystal symmetry and are generally considered to be isomorphous. There are actually three groups of alums because of slight differences in internal crystalline structure. **isomorphous** (*adj*).

exhibit (*v*) when a property, which is not always open to observation, becomes observable, then that property is said to be exhibited. Tin has three possible oxidation states, as an atom, as an ion with an electrovalency of 2, and as an ion with an electrovalency of 4; redox processes can change the electrovalencies. When tin forms tin (II) chloride, it exhibits an electrovalency of 2. **exhibition** (*n*).

vapour pressure/temperature curves for the allotropes of phosphorus (typical for a monotropic change)
m.p. of white phosphorus is T_1
m.p. of red phosphorus is T_2
rapid cooling forms white phosphorus slow cooling forms red phosphorus

monotropy

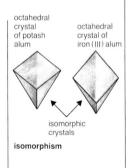

octahedral crystal of potash alum

octahedral crystal of iron (III) alum

isomorphic crystals

isomorphism

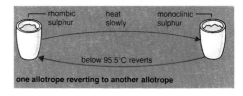

one allotrope reverting to another allotrope

revert (v) to go back to the previous state,
e.g. monoclinic sulphur, formed from rhombic sulphur,
when cooled below the transition temperature (p.13)
slowly reverts to rhombic sulphur, which is the stable
form at room temperature. Contrast **return**, which is to
go back to a previous position. An electron *returns* to its
original shell and *reverts* to its original energy level.

metastable (*adj*) describes a state, or equilibrium, which
is capable of a spontaneous (↓) change, but does not
undergo the change. The process of change is initiated
by a catalyst, or some outside impulse, or may take
place extremely slowly, e.g. a supercooled liquid
persisting below its freezing point is metastable; a
supersaturated solution is metastable, a crystal will start
the process of crystallization; white phosphorus is
metastable, it changes extremely slowly to red
phosphorus, the process can be speeded up by a
catalyst.

stable (*adj*) describes a state, or equilibrium, which
does not undergo change under the stated conditions.
Describes a substance which does not undergo a
physical change. *See* **labile**. **stability** (*n*).

thermostable (*adj*) describes a compound which does
not change physically or lose its properties when
heated.

spontaneous (*adj*) describes an event that takes place
without any external cause.

instantaneous (*adj*) describes an event that takes place
so quickly that the period of the event is not normally
observable, e.g. precipitation caused by two ions in
contact in solution is considered to be instantaneous,
as the period of the event is about 10^{-4} second. Very
sensitive apparatus can detect this period, but that is
not normal observation. **instant** (*n*).

transient (*adv*) describes an existence which lasts for a
very short period of time, e.g. a free radical has a
transient existence of about a thousandth of a second.

element (*n*) a substance (↓) consisting of atoms (p.29) all of which have a nucleus containing the same number of protons, i.e. the nuclear charge of all the atoms is the same. **elementary** (*adj*).

compound (*n*) a substance (↓) consisting of elements (p.15) chemically combined in a known constitution, with definite properties, and with an established formula.

substance (*n*) any type of matter (↓) that can be named and has definite, invariable properties. It can be an element, or a compound (↑), or it can have a known chemical constitution, but its formula is unknown or too complex to describe, e.g. many proteins. Before identification, a substance may be unknown; recognition of its properties can lead to an identification. Copper, oxyhaemoglobin and sodium chloride are substances.

material (*n*) any type of matter (↓) that can be named through recognition of its properties, although those properties may vary within limits, e.g. paper as the properties of paper vary; rubber is a material. **material** (*adj*).

matter (*n*) that which occupies space, exists in three states, and is observable. All matter consists of atoms (p.29). Matter is measured by (a) its mass; (b) its volume; (c) the amount of substance (↑) (not for materials).

mineral (*n*) a substance obtained from the earth; it has a definite chemical composition and crystalline structure, and recognizable physical and chemical properties. Examples are: (a) native elements, e.g. gold, (b) copper pyrites, (c) quartz. **mineral** (*adj*).

ore (*n*) a mineral (↑) from which a useful substance can be obtained. Ores generally have metals extracted from them, e.g. copper is extracted from copper pyrites ore.

binary compound a compound consisting of only two elements. Their stoichiometry include (a) AB; (b) AB_2; (c) AB_3; (d) A_2B_3; (e) AB_4; (f) AB_5; (g) A_2B_5.

ternary compound a compound consisting of three elements. The general formula of such compounds is $A_xB_yC_z$.

species (*n*) chemical species are any forms of matter taking part in a chemical reaction. They include atoms, activated atoms, molecules, activated molecules, ions (p.124) and free radicals (↓).

free radical an atom or group of atoms with one or more unpaired electrons. Many free radicals are only intermediate products in a chemical reaction, e.g. CH_3. Some free radicals, e.g. NO and CO, are stable.

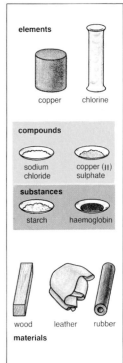

elements

copper   chlorine

compounds

sodium   copper (II)
chloride   sulphate

substances

starch   haemoglobin

wood   leather   rubber

materials

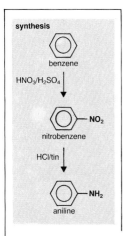

synthesis

benzene

HNO_3/H_2SO_4

NO_2
nitrobenzene

HCl/tin

NH_2
aniline

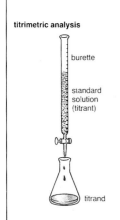

titrimetric analysis

burette

standard
solution
(titrant)

titrand

technique (*n*) a combination of knowledge, skill and
ability which enable a person to carry out a particular
activity.

analysis (*n*) one or more consecutive processes carried
out in order to establish the nature, composition or
constitution of a substance, mixture or compound,
e.g. the analysis of foods to determine the vitamin
content; the analysis of propane to determine its
chemical composition. **analyse** (*v*), **analyst** (*n*),
analytical (*adj*).

synthesis (*n*) the formation of a particular compound
either from its elements or from simpler compounds as
a result of chemical processes, e.g. the synthesis of
aniline from benzene; the synthesis of rubber from buta,
−1, 3–diene.

gravimetric analysis an analytical process based upon
the formation, isolation and weighing of precipitates
obtained in proportion by chemical reaction on the
substance being determined **gravimetry** (*n*).

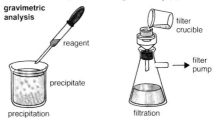

gravimetric
analysis

reagent

filter
crucible

filter
pump

precipitate

precipitation   filtration

thermogravimetric analysis the measurement of
changes occurring in the mass of substances as they
are steadily heated at a predetermined rate as a basis
for identifying them or for establishing their chemical
composition. **thermogravimetry** (*n*), **thermogram** (*n*).

trace analysis any analytical process carried out to
measure a very small amount of a substance,
compound or element (p.15). It usually applies to the
determination of concentrations in parts per million.

titrimetric analysis an analytical process in which
carefully measured solutions of known concentrations
are slowly added one to the other to establish how
much of the first will react with a known quantity of the
other. Acid-alkali, reduction-oxidation and
complexometric reactions are all used in titrimetric
analysis. **titrimetry** (*n*).

volumetric analysis *see* **titrimetric analysis**.

primary standard a chemical of very high purity, stable in air, soluble in water, and of a high relative molecular mass which can be employed in chemical analysis (p.17) as a reference material which will undergo reproducible, stoichiometric reactions, e.g. potassium hydrogen phthalate, benzoic acid.

acidimetric standard a primary standard (↑) with acidic characteristics suitable for acid-base titrations (p.137), e.g. constant boiling point hydrochloric acid.

alkalimetric standard a primary standard (↑) with basic characteristics suitable for acid-base titrations (p.137), e.g. sodium carbonate.

oxidation standard a primary standard (↑) used as an oxidising agent in redox titrations, e.g. potassium dichromate.

reduction standard a reducing agent suitable for use as a primary standard (↑) in redox titrations, e.g. pure iron.

iodine standard a solution of a known strength of iodine suitable for use in iodometric titrations. This is usually prepared in a solution of potassium iodide and the active species in the titration is the tri-iodide ion I3.

$$I_2 + I^- \rightleftharpoons I_3^+$$

iodine standard

argentometric standard a solution of a known strength of silver nitrate suitable for use in precipitation titrations (p.137) for the determination of halogens.

complexometric standard a solution of a known strength of a complexone (p.208), e.g. EDTA (p.208), suitable for use in complexometric titrations for the determination of metal ions in solution.

colorimetry (*n*) the study of the absorption and transmission of visible light by substances usually used as a basis for identification and quantification. As this area of study now involves ultraviolet and infrared radiation it is more common to use the general term spectrometry.

spectrometry (*n*) *see* **colorimetry** (↑).

absorption (*n*) the process by which a substance takes in electromagnetic radiation of particular wavelengths

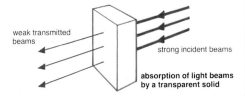

weak transmitted beams

strong incident beams

absorption of light beams by a transparent solid

thus increasing the energy in the chemical bonds (p.69) of the molecules (p.29). The quantity of the radiation either absorbed or transmitted can be measured and a characteristic spectrum for the substance under examination obtained. **absorbent** (*adj*)

sorption (*n*) an inclusive term for processes in which a substance takes into itself radiation, energy, gases or other substances by increasing its own energy state or by forming chemical bonds with the sorbed substances.

absorptiometer (*n*) a general name for any instrument which can be used to measure the absorbtion (↑) of visible light by a substance.

tintometer (*n*) a comparator (↓) in which chemical concentrations are determined by the production of characteristic intensities of ranges of colours which are compared with a series of coloured glass standards held in a plastic disc and observed through viewing holes in a plastic box.

tintometer comparator

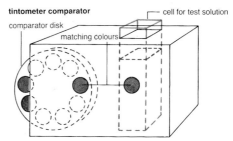

comparator disk

matching colours

cell for test solution

comparator (*n*) any scientific instrument in which concentrations of chemical compounds can be determined by comparing the intensities of colours they produce with a series of standards.

colorimeter (*n*) an instrument designed to measure the intensities of selected wavelengths of light transmitted or absorbed by substances. The wavelength of light selected is determined by filters or prisms by a process called monochromation.

spectrophotometer (*n*) a more advanced form of colorimeter (↑) measuring ultraviolet, visible and infrared radiation in which selection of individual wavelengths (monochromation) is carried out by rotating prisms or diffraction gratings and can be used to produce a complete absorption (↑) spectrum of the substance studied.

crystallization (*n*) the formation of crystals from a solution of a substance capable of existing in a crystalline form. Crystals are usually created if a solution is cooled to such an extent that it is super-saturated. The process may be used to assist in the purification of impure materials. **crystallize** (*v*).

mother liquor the term used for the residual solution after crystallization (↑) has occurred and the crystals filtered off. Partial evaporation of the mother liquor and further cooling often enable a second crop of crystals to be obtained from the solution.

recrystallization (*n*) the process of purifying a substance by crystallization (↑) and then redissolving the crystals in fresh hot solvent in order to repeat the crystallization. This may be carried out several times to improve progressively the purity of the chemical. **recrystallize** (*v*).

fractional crystallization the separation of two crystalline substances based upon differences in their respective solubilities. The mixture of the two materials is subjected to a series of recrystallizations (↑), such that after each process the mother liquor (↑) contains an increased proportion of one substance and the crystals an increased proportion of the other. After several repeated steps pure specimens of the two compounds are isolated.

fractional distillation the separation of mixtures of liquids with very similar boiling points by distilling them in an apparatus which includes a tall, vertical column (a reflux column) possessing a large area for condensing the vapours of the boiling liquids. The vapours have to be hot enough to ascend the column completely before they can be condensed and collected. The lower boiling liquids will be distilled off first and the distillate is collected in a series of fractions selected according to the boiling temperature measured at the head of the column. Fractional distillation forms the basis of many industrial processes of separation and purification, particularly for separating the components of crude oil.

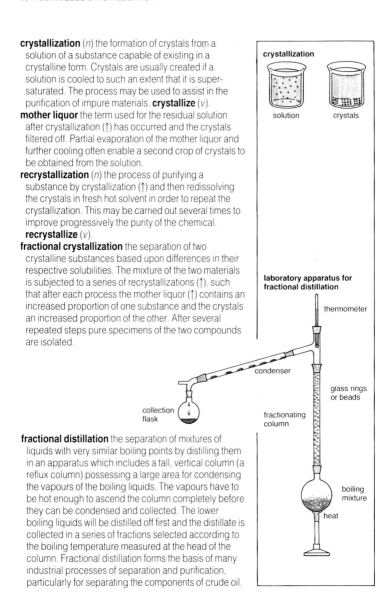

crystallization

solution   crystals

laboratory apparatus for
fractional distillation

thermometer

condenser

glass rings
or beads

collection
flask

fractionating
column

boiling
mixture

heat

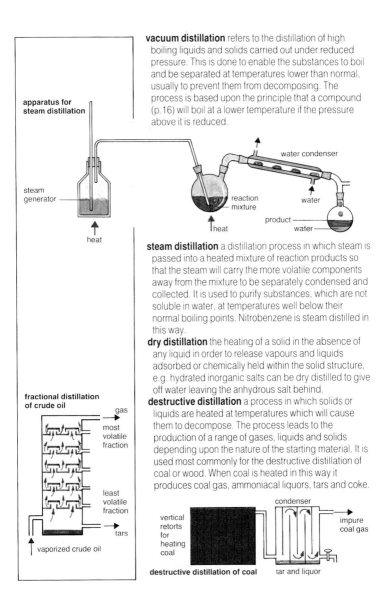

apparatus for steam distillation

vacuum distillation refers to the distillation of high boiling liquids and solids carried out under reduced pressure. This is done to enable the substances to boil and be separated at temperatures lower than normal, usually to prevent them from decomposing. The process is based upon the principle that a compound (p.16) will boil at a lower temperature if the pressure above it is reduced.

water condenser

steam generator

reaction mixture

water

product

water

heat

heat

steam distillation a distillation process in which steam is passed into a heated mixture of reaction products so that the steam will carry the more volatile components away from the mixture to be separately condensed and collected. It is used to purify substances, which are not soluble in water, at temperatures well below their normal boiling points. Nitrobenzene is steam distilled in this way.

dry distillation the heating of a solid in the absence of any liquid in order to release vapours and liquids adsorbed or chemically held within the solid structure, e.g. hydrated inorganic salts can be dry distilled to give off water leaving the anhydrous salt behind.

destructive distillation a process in which solids or liquids are heated at temperatures which will cause them to decompose. The process leads to the production of a range of gases, liquids and solids depending upon the nature of the starting material. It is used most commonly for the destructive distillation of coal or wood. When coal is heated in this way it produces coal gas, ammoniacal liquors, tars and coke.

fractional distillation of crude oil

gas

most volatile fraction

least volatile fraction

tars

vaporized crude oil

condenser

vertical retorts for heating coal

impure coal gas

destructive distillation of coal

tar and liquor

solvent extraction the use of a second immiscible
solvent to remove a compound already dissolved in
another solvent. The process is carried out by shaking
the second solvent with the solution in a special
separating funnel from which a lower layer of solvent or
solution can be run off. The solute initially in the first
solvent distributes itself between the two immiscible
solvents in a proportion dependent upon its distribution
ratio (↓). The two solvents must be immiscible for the
process to work, e.g. diethyl ether and water, water and
trichloromethane (chloroform).

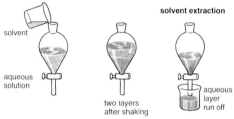

solvent

aqueous
solution

solvent extraction

two layers
after shaking

aqueous
layer
run off

distribution ratio the relative proportions in which a
substance will distribute itself between equal volumes
of two immiscible solvents when it is soluble in both,
e.g. the distribution ratio of iodine between water and
tetrachloromethane is 1:1.2. This ratio is a constant for
equal volumes of solvents at all concentrations of the
solute so long as neither solvent is saturated.

countercurrent distribution a process in which two
substances with similar distribution ratios (↑) are
separated as a result of carrying out multiple solvent
extractions in an apparatus (the Craig machine) which
continuously separates and re-extracts the individual
solvent layers from each extraction. The result is that
one substance becomes concentrated in the upper
solvent layers whilst the other is concentrated in the
lower solvent layer.

moving phase *see* **mobile phase**.

mobile phase the gaseous or liquid phase in
chromatography (↓) which moves over or through the
stationary phase and transports the components of the
mixture for separation.

stationary phase the solid particles or the viscous liquid
retained on the solid which act as the obstacle course
for molecules in chromatography (↓).

ascending paper chromatography

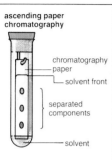

chromatography paper

solvent front

separated components

solvent

descending paper chromatography

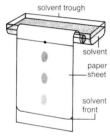

solvent trough

solvent

paper sheet

solvent front

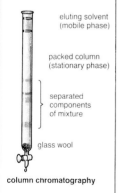

eluting solvent (mobile phase)

packed column (stationary phase)

separated components of mixture

glass wool

column chromatography

chromatography a term used for a range of processes in which mixtures of substances are separated as a result of their interactions with a gaseous or liquid mobile phase (↑) which transports the mixture over or through a liquid or solid stationary phase (↑). The process may be carried out in columns or on flat surfaces.

paper chromatography a process in which a small spot of a mixture of solutes is placed on an adsorbent paper strip which has one edge dipped into a solvent. Passage of the solvent carries the solutes along the paper and they are separated due to their different characteristics.

ascending paper chromatography a form of paper chromatography (↑) in which the lower edge of the paper strip is dipped in the solvent mobile phase (↑) which then ascends the paper by capillary attraction.

descending paper chromatography paper chromatography (↑) in which the upper edge of the paper strip is immersed in solvent in a trough so that the solvent phase can travel down the strip by a combination of capillary attraction and gravitational force.

column chromatography refers to any form of chromatographic separation involving the passage of a liquid or gaseous mobile phase (↑) over a solid surface contained within a glass or metal column. Traditional column chromatography was the separation of substances in solutions transported down glass columns packed with solid stationary phases (↑).

thin layer chromatography describes any chromatographic process in which the stationary phase (↑) is in the form of a thin layer spread uniformly over a glass, plastic or metal sheet along which the solvent mobile phase (↑) can ascend. It is used mainly for separations of small quantities of mixtures by adsorption chromatography (↓).

adsorption chromatography chromatographic processes in which separations are dependent upon the extent to which solutes in the gaseous or liquid mobile phases (↑) are adsorbed on solid surfaces of the stationary phases over which they pass.

partition chromatography refers to chromatographic processes in which substances are separated as a result of the different ratios with which they will distribute themselves between a stationary liquid phase and either a liquid or gaseous mobile phase. The success of the separations depends upon the solutes possessing different distribution ratios.

ion exchange chromatography describes separations
of mixtures carried out by chromatographic processes
as a result of the exchange of ions (p.124) in the mobile
liquid phase with ions held on the surface of the solid
stationary phase (p.22).

high performance liquid chromatography is an
inclusive term for chromatographic separations carried
out with liquid mobile phases (p.22) which are pumped
under pressure through stationary phases (p.22)
contained in capillary columns. The process can be
used for very small quantities of materials.

**block diagram of
apparatus for
gas chromatography**

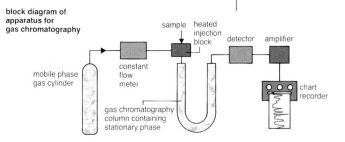

gas liquid chromatography describes chromatographic systems in which the mobile phase is a gas,
e.g. nitrogen or hydrogen, which carries the mixture to
be separated over a viscous stationary phase held
either on solid packing material or on the walls of the
column. Separation is achieved by partition of the
compounds between the gas and the liquid phase.

support (*n*) any solid material used to hold and retain a
liquid stationary phase (p.22) used in partition
chromatography (p.23), e.g. the walls of the tube itself
or inert powders, polymers or glass beads.

elution (*n*) a term referring to the process of passing a
mobile phase (p.22) over or through a stationary phase
(p.22) in order to transport solutes and to separate them
from each other. **elute** (*v*).

eluate (*n*) the solution of solutes in the mobile phase
(p.22) issuing from a chromatographic column after
separation has been carried out.

solvent front (*n*) the wet moving edge of solvent as it pro-
gresses along the paper strip in paper chromatography or
along the surface in thin layer chromatography (p.23) and
which can be used for the calculation of R_F values (↓).

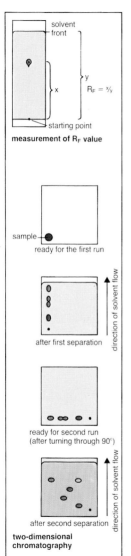

measurement of R_F value

sample

ready for the first run

after first separation

direction of solvent flow

ready for second run
(after turning through 90°)

direction of solvent flow

after second separation

two-dimensional
chromatography

R_F value the ratio of the distance moved by a particular compound (p.16) relative to the distance moved by the solvent front (↑) in chromatographic separations. It is used especially in paper chromatography (p.23) and thin layer chromatography (p.23) as a basis for identifying compounds as the ratio is a constant under specified conditions.

R_X value the ratio of the distance moved by a compound (p.16) with respect to the corresponding distance moved by a reference substance (X) under the same chromatographic conditions, e.g. the chromatographic separation of colours and dyes may be carried out with reference to the movement of the dye Butter Yellow using paper chromatography (p.23) or thin layer chromatography (p.23).

distribution coefficient *see* **distribution ratio** (p.22).

chromatoplate (*n*) refers to a glass or plastic plate or thin metal sheet covered with a layer of adsorbent, e.g. silica gel or alumina, suitable for use in thin layer chromatography (p.23).

chromatogram (*n*) a general term applied to a chromatographic record of a separation, e.g. the separated spots on a thin layer plate or from paper chromatography (p.23), or the automatically recorded peaks on a chart record obtained from gas chromatography (↑).

two-dimensional chromatography the process of running either a paper or thin layer chromatogram (↑) successively in two different solvent systems at right angles to each other in order to produce a separation over an area rather than in a straight line. To do this, the sample is placed as a small spot in one corner of the sheet or plate and the chromatogram run using the first solvent system. It is then dried before being rotated through 90° and then run using the second solvent system.

location (*n*) any process employed to establish the positions of solutes in column, paper and thin layer chromatography (p.23). This may be done by spraying the surface with chemical reagents to produce coloured zones, or by exposure to ultraviolet light.

development (*n*) the production of coloured derivatives of solutes separated by paper chromatography (p.23) or thin layer chromatography (p.23) by spraying with selective reagents, used to establish the location (↑) of the individual substances.

activation (*n*) a procedure used in adsorption chromat-ography (p.23) to improve the ability of the stationary phase to retain or slow down the movement of solutes being separated, e.g. alumina and silica gel in thin layer chromatography are activated by heating them to remove any adsorbed moisture. **activate** (*v*).

deactivation (*n*) stationary phases (p.22) in adsorption chromatography (p.23) lose their ability to retain solutes in chromatographic separations if they absorb moisture or get wet. They are said to be deactivated. Polar adsorbents and support (p.24) materials may also be deliberately deactivated by reacting their polar groups with compounds which possess non-polar groups such as silicones.

aeration (*n*) the process of bubbling or passing air through a liquid either to increase the amount of air dissolved in the liquid, to oxidize impurities or to assist in flotation (↓) processes. **aerate** (*v*), **aerator** (*n*).

froth flotation the separation of useful minerals from their ores by aeration (↑)of a mixture of the ground ore in water or oil to which a frothing agent (↓) has been added. The mineral particles are carried up on the bubbles whilst the gangue settles out.

oil flotation a froth flotation (↑) process in which an oil is added to form an insoluble upper layer to the flotation tank to assist in the separation of minerals with particular densities.

frothing agent a substance, such as soap or detergent which acts by lowering the surface tension at the air-liquid interface in froth flotation (↑) processes. It assists in the formation of the bubbles creating the froth.

collector (*n*) a chemical compound added to froth flotation processes to assist in the flotation of the mineral particles by attaching themselves to the surface of the particles. Such compounds include long chain alcohols and detergents.

activator (*n*) an inorganic ion used to assist the attachment of the collector (↑) to the particle being separated by froth flotation (↑), e.g. sodium sulphide is of particular assistance with lead ores.

dichroic (*adj*) describes crystals which have different colours when observed from different directions. A dichroic crystal absorbs light in one plane more than in a plane at right angles; if the crystal is sufficiently thick, it will absorb sufficient light in one plane to produce plane polarized light in the plane at right angles.

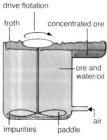

drive flotation

froth | concentrated ore

ore and water/oil

air

impurities | paddle

froth flotation

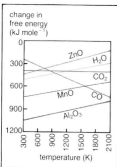

Ellingham diagram for oxides

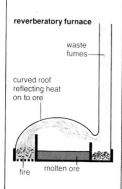

reverberatory furnace

waste fumes

curved roof reflecting heat on to ore

fire   molten ore

Ellingham diagram refers to a graphical representation of the variation of the free energy (p.186) of formation of a particular class of compounds, e.g. oxides, with respect to temperature. Essentially the diagrams indicate which metals and their compounds are likely to interact with each other and the suitability of using such substances as carbon and carbon monoxide for the reduction of metal oxides to the pure free metals.

smelt (v) the process of separating a metal from an ore by heating the crushed ore with a substance which acts as a reducing agent, e.g. coke, at a temperature which will melt the metal. The impurities from the process float as slag on the surface of the molten metal.

flux (n) refers to a chemical added to a metal or alloy to help it flow and to promote fusion, e.g. a flux is used to help solder to form a clean union between a wire and a terminal.

slag (n) a fused substance obtained as waste from the reduction of metal ores. It is obtained mainly as calcium silicate formed by interaction between the ore waste and limestone added to the heated mixture.

fuse (v) to melt substances or objects in order that they will blend together into a single unified mass.

calcine (v) (1) to drive off volatile components from a solid by heating to a high temperature. (2) to form a metal oxide by heating the metal at a high temperature in air or oxygen.

sinter (v) to heat solid particles at a temperature slightly below their melting point to an extent that they will coalesce to form a single porous mass. Glass, ceramics and metals can be sintered.

pyrolysis (n) the process of decomposing substances by heating them, usually in a limited supply of air or oxygen. It leads to the breakdown of the larger molecules and the formation of more simple chemical compounds. **pyrolyse** (v), **pyrolyser** (n).

degradation (n) refers to the breaking down of a substance to simpler chemical compounds usually in a series of defined processes. **degrade** (v).

concentrate (v) to remove solvent from a solution so that the original amount of solute is in a lesser quantity of solvent. **concentration** (n).

reverberatory furnace a furnace for smelting ores in which heat from the fire is directed down by means of a low curved roof on to the hearth covered with ore and reactants.

open-hearth furnace a furnace in which iron ore and pig-iron are smelted in a shallow bowl-shaped hearth lined with limestone. Heated air is passed over the liquid metal surface to oxidise the impurities in the iron.

wedge roaster a reverberatory furnace with mechanical rabbles used for roasting ores; it expels unwanted volatile components, e.g. sulphur and arsenic, as they are released during the smelting process. The wedge roaster consists of a series of circular multiple hearths with rotating, water-cooled rabbles extending from a central shaft which spread the ore out and sweep it downwards to produce a thin layer of material with a large surface area.

carbonization (*n*) refers to any process used for producing a fixed concentration of carbon in steel.

cementation (*n*) an old process used for the manufacture of steel in which bars of wrought iron were heated in red-hot charcoal for several days.

annealing (*n*) a process used to relieve stresses in metals by heating the metal and then allowing it to cool down at a slow, regulated, steady rate.

quenching (*n*) refers to the hardening of steel by heating it until it is red hot then rapidly cooling it by plunging it into cold water or sometimes cold oil.

tempering (*n*) a process for introducing a specified degree of hardness in steel. Steel which has been quenched (↑) is reheated to a predetermined temperature at which it is maintained for a short period before being allowed to cool. This reduces the hardness but increases its elasticity.

age-hardening (*n*) a process by which the hardness of a metal or alloy increases with time after it has been refined.

work-hardening (*n*) another name for strain-hardening in which a metal or alloy is hardened by rolling or stretching it. This 'working' of the metal leads to dislocations of the crystalline structure.

plant (*n*) a collective word describing the reaction vessels, furnaces, heaters, stirrers, pipes and equipment used together as an integrated unit in order to carry out an industrial process.

recycling (*n*) describes any procedure in which waste materials are either used for a different purpose or are re-processed to produce new materials, e.g. scrap iron is recycled to produce new steel; old glass bottles may be recycled to give new glass.

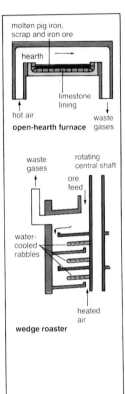

molten pig iron, scrap and iron ore

hearth

limestone lining

hot air   waste gases

open-hearth furnace

waste gases   rotating central shaft

ore feed

water-cooled rabbles

heated air

wedge roaster

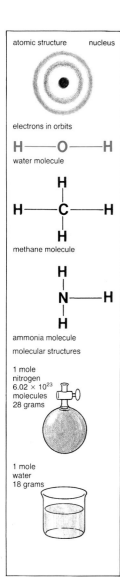

atomic structure   nucleus

electrons in orbits

H———O———H

water molecule

methane molecule

ammonia molecule

molecular structures

1 mole
nitrogen
6.02×10^{23}
molecules
28 grams

1 mole
water
18 grams

atom (*n*) the simplest portion of an element (p.16) which possesses the properties of that element and can take part in chemical reactions to form molecules (↓). An atom is formed from a central nucleus containing neutrons (p.44) which are uncharged, protons (p.44) which possess positive charges, and negatively charged electrons (p.43) which move in orbitals (p.49) around the nucleus.

molecule (*n*) refers to a group of atoms (↑) joined together by chemical bonds (p.69) to form the smallest recognizable particle of a chemical compound or of an element in a free state, e.g. the bromine molecule consists of two bromine atoms combined together as Br_2, the methane molecule consists of four hydrogen atoms joined to a carbon atom as CH_4.

relative atomic mass the ratio of the average mass of the atoms of an element to one-twelfth of the mass of the ^{12}C nuclide. The value is dimensionless.

relative molecular mass the ratio of the average mass of the molecules of a chemical compound to one-twelfth of the mass of the ^{12}C nuclide. The value is equal to the sum of the relative atomic masses (↑) of the atoms forming the molecules.

relative formula mass the mass of one mole (↓) of a chemical compound (p.16) or element (p.16) expressed in grams, calculated from the relative atomic masses (↑) for each of the atoms in the formula, e.g. the relative formula mass for methane, CH_4, is 16 g, for ammonia, NH_3, it is 17 g, and for benzene, C_6H_6, it is 78 g.

amount (*n*) an unspecified quantity of a substance referring to a mass or volume of a solid, liquid or gas.

mole (*n*) the standard term for the physical quantity of an amount of substance. A mole of any substance contains 6.02×10^{23} particles of that substance, whether atoms (↑), ions (p.124), molecules (↑) or electrons (p.43). The amount of a substance may be given in moles, thus 1 mole of ammonia has a mass of 17 g and contains 6.02×10^{23} molecules of ammonia. **molar** (*adj*).

mole fraction refers to an amount of substance less than a mole (↑) expressed as a decimal quantity, e.g. 0.35 mole is a mole fraction; 4 g of sodium hydroxide in 1 dm³ of solution is a (4/40) 0.1 M solution.

molar mass the mass of one mole (↑) of a substance expressed in grams, e.g. molar mass of water H_2O is 18 grams; molar mass of ethanol C_2H_5OH is 46 grams.

molar volume the volume occupied by one mole of a substance under specified conditions. For all gases measured at s.t.p the molar volume is a constant of $22.4\,dm^3$ and is indicated by the symbol V_m.

Avogadro constant refers to the number of particles in a mole (p.29). This is a constant number of 6.02×10^{23} particles. The symbols L or N_A are used to indicate the constant.

concentration (n) the amount (p.29) of one substance dissolved in another, usually applied to the quantity of the substance dissolved in a liquid to give a solution. Concentration may be expressed in a variety of ways, e.g. as grams of solute in $1\,dm^3$ of solution (weight/volume); as volume of solute in $1\,dm^3$ of solution (volume/volume); as the weight of solute in $1\,kg$ of solution (weight/weight). The values are often expressed as percentages, e.g. 98 grams of sulphuric acid in $1\,dm^3$ of solution has a concentration of $98\,g\,dm^{-3}$, $1\,mol\,dm^{-3}$ or 9.8% *W/V*. **concentrate** (v), **concentrated** (*adj*).

molar concentration refers to the amount of substance, moles (p.29) or molar fraction (p.29), dissolved in a specified volume of solution, e.g. 0.7 moles of sodium hydroxide in $250\,cm^3$ of solution.

molarity (n) a term used to refer to the molar concentration (↑) in moles per dm^3 for solutions used especially for analytical chemistry involving titrimetry (p.137) and gravimetry (p.17).

molality (n) the number of moles (p.29) of a substance dissolved in $1\,kg$ of solvent.

M-value (n) an abbreviated form of indicating the molarity (↑) of a solution in terms of the number of moles (p.29) of solute dissolved in $1\,dm^3$ of solution, e.g. if a solution has a molarity of 0.7 it is expressed as being $0.7\,M$.

stoichiometric compound a compound which can be represented by a whole number chemical formula and obeys the laws of constant chemical composition, e.g. methane CH_4 is a stoichiometric compound. **stoichiometry** (n).

non-stoichiometric compound refers to a chemical compound which does not always have a constant chemical composition and cannot be readily represented by a whole number chemical formula, (↓) e.g. titanium oxides are non-stoichiometric with formulae of $TiO_{1.9}$ and $TiO_{1.65}$.

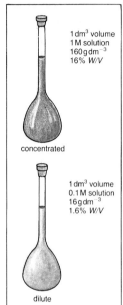

$1\,dm^3$ volume
$1\,M$ solution
$160\,g\,dm^{-3}$
16% *W/V*

concentrated

$1\,dm^3$ volume
$0.1\,M$ solution
$16\,g\,dm^{-3}$
1.6% *W/V*

dilute

concentration and dilution
copper
sulphate
solutions

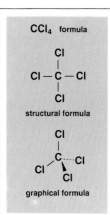

CCl₄ formula

structural formula

graphical formula

carbon tetrachloride

structural isomerism
structural isomers of
nitrotoluene

formula (*n*) (1) a combination of chemical symbols used to show the number of atoms (p.29) of each element combined in the structure of a molecule (p.29) or ion (p.124) of a chemical compound (p.16). This may be in the form of a structural formula (↓) that shows the order in which the atoms are joined to each other, or a graphical formula (↓) in which a three-dimensional representation of the structure is presented. (2) a presentation of mathematical or physical symbols used to indicate interrelationships between mathematical quantities or physical properties, e.g. $A = \pi r^2$ is the formula for the area of a circle. **formulate** (*v*).

empirical formula the simplest formula for a compound which indicates the atoms in their lowest proportions to each other, e.g. ethane has a molecular formula C_2H_6 but its empirical formula is CH_3.

molecular formula a formula (↑) of a chemical, arranged to give the number of atoms of each element present in the structure of the molecules of that substance, e.g. $C_2H_4O_2$ shows that each molecule (p.29) of ethanoic acid contains two atoms of carbon, four atoms of hydrogen and two atoms of oxygen. The manner in which these are arranged is shown by a structural formula (↓) or by a graphical formula (↓).

structural formula a chemical formula written in a form that shows the order and arrangement of the atoms in the molecule, e.g. the structural formula for methanol is CH_3OH, and that for ethanoic acid is CH_3COOH.

graphical formula a chemical formula (↑) drawn to give a three-dimensional representation of a molecular structure showing how the atoms are arranged with respect to each other. The best means of doing this is by what is known as the line and wedge method; a dotted line shows the position of a group (p.51) below the plane of the paper, and a wedge-shaped bond indicates a group which is above the plane of the paper.

isomerism (*n*) refers to chemical compounds possessing the same chemical formula (↑) but different chemical structures, e.g. propanol and ethoxymethane are isomeric as they both have the formula C_3H_8O. **isomeric** (*adj*), **isomerize** (*v*).

structural isomerism isomerism (↑) exhibited by chemical compounds (p.16) which possess the same chemical formula (↑) but different structural formulae. Such substances may be chemically similar, as with the dimethyl benzenes (xylenes), or dissimilar.

stereoisomerism (*n*) isomerism (p.31) arising from molecules with an identical structural formulae but a different spatial arrangement of the atoms in the molecule (p.29). Graphical formulae (p.31) are used to illustrate the difference between **stereoisomers**. Most stereoisomerism arises in organic compounds due to the tetrahedral arrangement of the four covalent bonds (p.70) of the carbon atoms. Two forms of stereoisomerism exist: **optical isomerism** (↓) and **geometrical isomerism** (↓).

optical isomerism refers to the isomerism of a compound which does not possess a structure which can be superimposed upon that of its mirror image. Although the two compounds are chemically identical in every way, there is a different spatial arrangement of the atoms in the molecule. If polarized light is passed through separate solutions of the two isomers the plane of the polarization is turned to the same extent, but in opposite directions, by the two solutions.

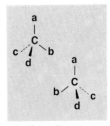

optical isomerism

enantiomorphs (*n*) (1) two or more isomeric (p.31) molecules which exhibit optical isomerism (↑). (2) two crystal structures which are mirror images of each other.

chiral (*n*) refers to an atom (p.29), usually carbon, which imparts optical isomerism (↑) to a compound (p.16) by virtue of the chemical groups (p.51) to which it is bonded. If a carbon atom is joined to four different groups then it is referred to as an asymmetric carbon atom or a **chiral centre** and the substance will be optically active.

racemic mixture a mixture of equal quantities of enantiomorphs (↑) such that the optical activity of one is cancelled out by that of its mirror image. The net optical activity of a racemic mixture is zero.

racemate (*n*) a chemical compound (p.16) possessing two or more chiral (↑) centres such that one half of the molecule (p.29) is a mirror image of the other. As a result, it is optically inactive and does not rotate the plane of polarized light.

Cl   enantiomorphs   Cl   chiral centre

H — C — OH   HO — C — H

H_3C mirror images CH_3

chiral centre

COOH
|
H — C — OH
- - - - - - - - - - - - -
H — C — OH
|
COOH

racemic form of tartaric acid with two chiral centres

geometrical isomers

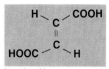

fumaric acid   trans

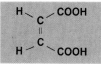

maleic acid   cis

geometrical isomerism a form of isomerism (p.31) occurring due to the spatial arrangement of atoms relative to a double bond (p.72) or a ring structure in a molecule (p.29). The double bond or ring system prevents free rotation occurring between the carbon atoms such that two substituents may both be arranged on one side of the bond (cis or syn orientation) or on opposite sides (trans or anti-syn orientation). The two isomers have different physical properties but may have different or similar chemical properties.

cis-form (n) a geometrical isomer (↑) in which two different groups (p.51) are on the same side of the double bond or ring.

trans-form (n) a geometrical isomer (↑) in which two groups in the molecule (p.29) are on opposite sides of the double bond or ring structure.

tautomerism (n) the existence of two interconvertible isomers in a state of dynamic equilibrium. In the most common form of tautomerism a hydrogen atom changes its position from being joined to a carbon atom (the keto-form) to an oxygen atom (the enol-form). Acetoacetone (pentan-2,4-dione) is a keto-form of a tautomer. **tautomer** (n), **tautomeric** (adj).

$$CH_3C\ CH_2C\ CH_3 \rightleftharpoons CH_3C = CHC\ CH_3$$

tautomerism
tautomeric forms   O   O   OH   O
of pentan−2,4−dione   keto form   enol form

keto-form (n) one of the two interconvertible tautomers in which the labile hydrogen atom is attached to a carbon atom, and in which there is a free carbonyl group (p.51).

enol-form (n) one of the two interconvertible tautomers in which the labile hydrogen atom is attached to an oxygen atom on a carbon atom adjacent to a carbon-carbon double bond (p.72).

chain structure any chemical structure in which a number of identical atoms, groups or repeating units are joined together one after the other. It refers especially to organic compounds formed from chains of carbon atoms. Chains may be straight or branched.

straight chain a chain of carbon atoms in which each atom is not joined to more than two other carbon atoms, such that a branched chain (p.34) is not possible.

straight chain
pentane $CH_3 — CH — CH — CH_2 — CH_3$

branched chain a chain of carbon atoms in which one or more carbon atoms are joined to at least three other carbon atoms, such that the structure is not one continuous straight chain (p.33).

ring structure any arrangement of atoms (p.29) joined together to form a ring. Carbon compounds may contain a variety of ring structures with double or single bonds between carbon atoms.

alicyclic ring a ring structure (↑) formed from carbon atoms joined to each other mainly by single bonds.

hydrocarbon (n) any organic chemical compound formed solely from carbon and hydrogen atoms (p.29) combined together. Hydrocarbons may be classified as aliphatic chain structures (p.33), alicyclic (↓) or aromatic (↓).

alkanes (n) straight chain (p.33) and branched chain (↑) hydrocarbons (↑) in which there are no double bonds. These are saturated hydrocarbons. The simplest alkane is methane CH_4. They possess the general formula C_nH_{2n+2} forming an homologous series (↓).

alkenes (n) straight chain (p.33) and branched chain (↑) hydrocarbons (↑) in which one or more of the carbon-carbon bonds is a double bond. These are unsaturated hydrocarbons and the simplest alkene is ethene (ethylene). They possess the general formula C_nH_{2n} forming an homologous series (↓).

alkynes (n) straight chain (p.33) and branched chain (↑) hydrocarbons (↑) in which one or more of the carbon-carbon bonds is a triple bond (p.68). These are unsaturated hydrocarbons and highly reactive. The simplest alkyne is ethyne (acetylene). They possess the general formula C_nH_{2n-2} forming an homologous series.

arenes (n) hydrocarbons in which the main structure is based on the six-membered ring system of benzene.

cycloalkanes (n) a group of saturated hydrocarbons (↑) possessing no carbon-carbon double or triple bonds in which the carbon atoms are joined together to form a closed ring.

aromatic (adj) describes a wide range of chemical compounds (p.16) which incorporate either a benzene ring, or a ring possessing similar characteristics.

aliphatic (adj) any organic compound (p.16) in which the main carbon structure is a straight chain (p.33) or branched chain (↑) hydrocarbon (↑).

alicyclic (adj) any organic compound in which the main structure or skeleton consists of carbon atoms formed in a ring joined to each other mainly by single bonds.

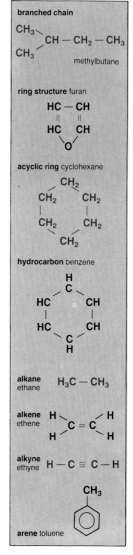

branched chain

CH_3
$\quad$ CH — CH_2 — CH_3
CH_3
$\qquad$ methylbutane

ring structure furan

HC — CH
$\ \parallel \quad \parallel$
HC $\quad$ CH
$\quad$ O

acyclic ring cyclohexane

$\qquad CH_2$
$CH_2 \qquad CH_2$
$\ | \qquad\qquad |$
$CH_2 \qquad CH_2$
$\qquad CH_2$

hydrocarbon benzene

$\qquad$ H
$\qquad$ C
HC $\qquad$ CH
$\ | \qquad\qquad |$
HC $\qquad$ CH
$\qquad$ C
$\qquad$ H

alkane $\quad H_3C$ — CH_3
ethane

alkene $\ $ H$\qquad$ H
ethene $\qquad$ C = C
$\quad$ H$\qquad\qquad$ H

alkyne $\ $ H — C ≡ C — H
ethyne

$\qquad CH_3$

arene toluene

cycloalkane cyclopropane

aromatic compound benzene

acyl
acyl groups

methanoic
(formic acid)

formyl

ethanoic
(acetic acid)

acetyl

alcohol	homologous series of alcohols $CnH_{2n+1}OH$	boiling point °C
methanol	CH_3OH	64
ethanol	CH_3CH_2OH	78
propanol	$CH_3CH_2CH_2OH$	97
butanol	$CH_3CH_2CH_2CH_2OH$	117
pentanol	$CH_3CH_2CH_2CH_2CH_2OH$	138
hexanol	$CH_3CH_2CH_2CH_2CH_2CH_2OH$	156

homologous series refers to closely related chemical compounds (p.16) which differ from each other by a constant change in the molecular composition, e.g. the alkanes (paraffins) have the general formula C_nH_{2n+2} and each member of the series differs from the previous one by an extra $(-CH_2-)$ unit. An homologous series of compounds may also possess the same functional group in the same position in the various molecules, as with the straight chain (p.33) alcohols (p.39). Within the homologous series the physical properties of the members change as the relative molecular masses (p.29) increase and they become more solid, with higher melting points and higher boiling points.

homologue (n) one member of an homologous series (↑), e.g. propanol is the third homologue of the homologous series of alcohols.

general formula a mathematical representation of the formula (p.31) of any member of an homologous series, e.g. the general formula for alkanes (↑) is C_nH_{2n+2} and that for alcohols is $C_nH_{2n+2}O$. The replacement of n in the formula by the appropriate number for the homologue (↑) in the series gives the formula of that compound.

alkyl (adj) refers to any organic group resulting from the removal of a hydrogen atom from an aliphatic (↑) compound, e.g. the methyl group (CH_3-); the ethyl group (C_2H_5-); the acetylenic group ($HC\equiv C-$). In an abbreviated form an alkyl group may be represented generally in chemical formulae (p.31) by R–. **alkylation** (n).

acyl (adj) refers to any organic group derived as a result of the loss of an hydroxyl group (p.36) from the functional group (p.36) of a carboxylic acid (p.36). **acylation** (n).

aryl (*adj*) describes the aromatic (p.34) group obtained when one atom of hydrogen is removed from benzene or other aromatic compounds. In an abbreviated form an aryl group may be represented in chemical formulae by Ph– (for phenyl) or by Ar– (for aromatic). **arylation** (*n*).

functional group an atom or a combined group of atoms which impart specific chemical properties to organic compounds. Substances with identical functional groups can form homologous series (p.35), and the physical and chemical properties of the series will change slightly as the size of the alkyl (p.35) or aryl (↑) system increases and the proportional effect of the functional group decreases. Common functional groups are: hydroxyl (↓), carboxyl (↓), amino (p.38), carbonyl (↓) and chloro.

hydroxyl group the main functional group (↑) found in alcohols (p.39) and phenols, imparting their unique properties to them. The hydroxyl group –OH is highly reactive and can be completely replaced by other functional groups; the hydrogen atom of the hydroxyl group may be replaced as a separate entity as it is acidic in character.

carbonyl group the main functional group (↑) found in aldehydes and ketones (p.39). It is also present as part of the amido group (p.38) and the carboxylic group (↓). The carbonyl group, consisting of a carbon atom joined to oxygen by a double bond, $C=O$, is reactive, and the oxygen atom can be replaced in condensation reactions, or one of the carbon-oxygen bonds broken in addition reactions. Carbonyl groups in aldehydes are more reactive than those in ketones.

carboxylic group the functional group (↑) associated with carboxylic acids and written in formulae (p.31) as –COOH. The hydrogen atom is ionizable and organic carboxylic acids are weakly acidic and capable of forming salts by neutralization, and esters (p.40) by reaction with alcohols (p.39).

aryl
phenyl group

molecule with
hydroxyl group
ethanol CH_3CH_2OH

molecule with
carbonyl group
benzophenone

carboxylic group

$H_3C - C {\overset{\displaystyle O}{\underset{\displaystyle OH}{\Big<}}}$

carboxylic acid
ethanoic (acetic) acid

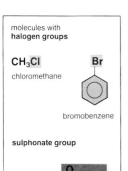

molecules with
halogen groups

CH₃Cl
chloromethane

Br

bromobenzene

sulphonate group

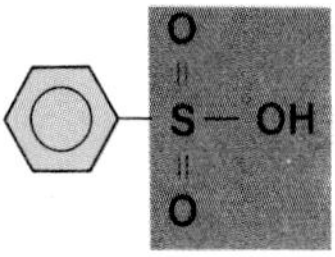

halogen group refers to functional groups (↑) in organic compounds created by the presence of the halogen atoms fluorine, chlorine, bromine or iodine. In alkyl (p.35) compounds these groups are easily replaced by other chemical groups and are useful as reactive centres in chemical synthesis. Aromatic halo-compounds are more stable and less reactive than are alkyl compounds. The reactivity of the halogen group increases in the order: F < Cl < Br < I. **halide** (n).

sulphonate group the $-SO_2OH$ functional group (↑) found mainly in aromatic (p.34) compounds. It is introduced into the benzene ring by reaction with concentrated sulphuric acid. The hydrogen atom is ionizable and the aromatic sulphonic acids can be neutralized to give salts. Long chain hydrocarbons with sulphonate groups attached to them are used as detergents (p.175).

nitro group the $-NO_2$ functional group (↑) introduced into aliphatic (p.34) and aromatic (p.34) compounds by nitration with concentrated nitric acid sometimes mixed with concentrated sulphuric acid. Aromatic nitro compounds are useful chemical intermediates as they can be reduced through several steps to amines (p.38). The nitro group in aliphatic compounds is tautomeric.

nitro group
tautomeric group

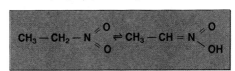

cyano group — C ≡ N

cyano group the $-CN$ functional group (↑) introduced into organic compounds by the action of sodium cyanide, NaCN, on alkyl halides (↑) to produce alkyl nitriles. When hydrolysed they initially form amides (p.38) which are further hydrolysed to carboxylic acids. Reduction of the cyano group produces the corresponding amines (p.38).

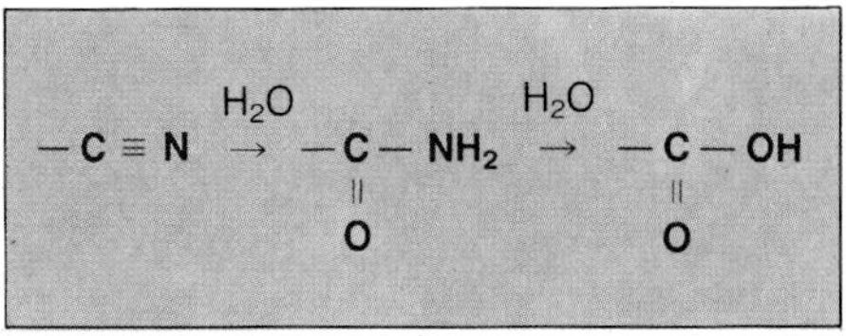

hydrolysis of a **cyano group** to a carboxylic acid

amino group the $-NH_2$ functional group (p.36) found in organic primary **amines**. Amines are formed by reduction of the cyano group (p.37) in organic nitrile compounds, or by reaction between ammonia and alkyl halides. Primary amines are produced by replacement of one hydrogen atom in ammonia, secondary amines by replacement of two hydrogen atoms and tertiary amines by replacement of all the hydrogen atoms. Short chain (p.33) amines are gases whilst longer chain amines are liquids. They form weakly basic solutions when dissolved in water and can form salts with inorganic acids.

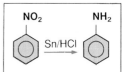

reduction of a nitro group to an **amino group**

$R - Cl + NH_3 \rightarrow RNH_2$ primary amine

$R - Cl + R'NH_2 \rightarrow R - \underset{\underset{R'}{|}}{N}H$ secondary amine

$R - Cl + R'_2NH \rightarrow R - \underset{\underset{R'}{|}}{N} - R'$ tertiary amine

formation of amines

amido group the $-CONH_2$ functional group (p.36) in organic compounds introduced either by partial hydrolysis of cyano groups (p.37) or by dehydration of the corresponding ammonium salt of the carboxylic acid. **amide** (*n*).

amido group
dehydration of ammonium salt to give an amide

$$CH_3C \overset{O}{\underset{ONH_4}{\diagup}}$$

$$\downarrow \ -H_2O$$

$$CH_3C \overset{O}{\underset{NH_2}{\diagup}}$$

azo group the highly reactive $-N=N-$ functional group (p.36) found in diazo compounds and produced by the action of nitrous acid on primary amines at temperatures in the range $0°$ to $-10°C$. Diazo compounds are intermediates in the manufacture of a range of dyes.

nomenclature (*n*) refers to the naming of chemical compounds. This may be by means of trivial (traditional) names or by a systematic procedure based mainly upon the length of the carbon chain (p.33) or the size of the carbon ring structure.

nomenclature

formula	trivial name	systematic name
CH_3OH	methyl alcohol	methanol
CH_3COCH_3	acetone	propan − 2 − one
$HC \equiv CH$	acetylene	ethyne
$H_2C = CH_2$	ethylene	ethene
$CH_3C \overset{O}{\underset{H}{\diagup}}$	acetaldehyde	ethanal

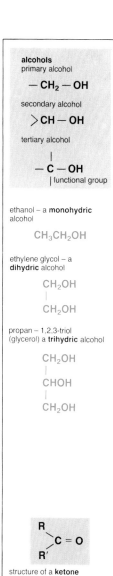

alcohols
primary alcohol

$$- CH_2 - OH$$

secondary alcohol

$$> CH - OH$$

tertiary alcohol

$$- C - OH$$
| functional group

ethanol – a **monohydric** alcohol

$$CH_3CH_2OH$$

ethylene glycol – a **dihydric** alcohol

$$CH_2OH$$
|
$$CH_2OH$$

propan – 1,2,3-triol (glycerol) a **trihydric** alcohol

$$CH_2OH$$
|
$$CHOH$$
|
$$CH_2OH$$

$$R$$
$$\diagdown$$
$$C = O$$
$$\diagup$$
$$R'$$

structure of a **ketone**

alcohols (*n*) organic compounds containing one or more hydroxyl groups in the molecule other than those connected directly to an aromatic (p.34) ring. Alcohols are formed by hydrolysis of alkyl halides (p.37) or esters (p.40) or by reduction of aldehydes (↓). They are classified according to the nature of the carbon atom to which the hydroxyl group is joined, as the reactivity of the alcohol is determined by the nature of the other groups joined to that carbon atom. Alcohols form esters when reacted with carboxylic acids (p.36), and can be oxidised successively to aldehydes or ketones (p.39) and then to carboxylic acids. The hydroxyl group can also be replaced by other functional groups (p.36), and the hydrogen atom can be replaced by alkali metals to produce alkoxides.

monohydric (*adj*) refers to an alcohol (↑) possessing a single hydroxyl group (p.36), e.g. ethanol.

dihydric (*adj*) describes an alcohol (↑) possessing two hydroxyl groups (p.36), e.g. ethan–1,2–diol (ethylene glycol).

trihydric (*adj*) describes an alcohol (↑) with three hydroxyl groups (p.36), e.g. propan–1,2,3–triol (glycerol).

aldehydes (*n*) organic compounds possessing the functional group –CHO in the molecule. Aldehydes may be prepared by partial oxidation of the corresponding primary alcohol (↑). They are named according to the length of the alkane (p.34) chain with the terminal *al* added, e.g. with four carbon atoms, the alkane is butane and the aldehyde is butanal. Reduction of aldehydes produces primary alcohols, and the carbonyl group will react with many substances, including sodium hydrogen sulphite, to form addition (p.212) compounds.

ketones (*n*) organic compounds (p.16) possessing a carbonyl group (p.36) on a carbon atom also attached to two alkyl (p.35) or aryl (p.36) groups. They may be prepared by partial oxidation of secondary alcohols. Ketones react in a similar manner to aldehydes (↑), forming addition (p.212) compounds and can be reduced to secondary alcohols (↑). They are named systematically according to the alkane name of the longest carbon chain within the ketone molecule with the carbonyl group carbon atom being numbered and the terminal 'one' added to the name, e.g. the ketone $CH_3CH_2COCH_2CH_3$ has five carbon atoms in the chain and is pentan–3–one.

ethers (*n*) describes organic compounds with two alkyl (p.35) or aryl (p.36) groups joined together by an oxygen atom (p.29). Short chain (p.33) ethers are highly volatile and flammable. They are used most commonly as solvents.

epoxides (*n*) cyclic ethers (↑) in which an oxygen atom forms a bridge between two adjacent carbon atoms. The best known member of the family is ethylene oxide which is used to give long chain polymers (p.236) by condensation reactions involving the breakage of one of the oxygen bridge bonds.

esters (*n*) organic compounds produced by substituting the hydrogen atom of a carboxylic (p.36) acid by an alkyl (p.35) or aryl (p.36) group. This is achieved by reacting the acid with an appropriate alcohol. The reaction is reversible as the ester can be hydrolysed back to the original acid and alcohol. Short chain (p.33) esters are sweet-smelling liquids which are used as solvents.

phenols (*n*) organic aromatic (p.34) compounds in which one or more hydroxyl groups are directly attached to the aromatic ring structure (p.34). They are acidic compounds and may be used as antiseptics. Most phenols are solids and possess strong characteristic odours. They are very important as starting points in chemical syntheses.

acid anhydrides organic compounds (p.16) formed by the removal of a molecule (p.29) of water from between two molecules of a carboxylic (p.36) acid. This forms a symmetrical anhydride. Mixed anhydrides are prepared by reacting the sodium salt of one carboxylic acid with the acid chloride of another.

acid anhydride

$$\begin{array}{c} RCOOH \\ RCOOH \end{array} \xrightarrow[\rightarrow]{-H_2O} \begin{array}{c} RCO \\ RCO \end{array}\!\!\!\diagdown O$$

symmetrical anhydride

$$\begin{array}{c} RCOCl \\ R'COONa \end{array} \xrightarrow[\rightarrow]{-NaCl} \begin{array}{c} RCO \\ R'CO \end{array}\!\!\!\diagdown O$$

mixed anhydride

nitriles (*n*) organic compounds containing the cyano group (p.37).

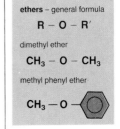

ethers – general formula

$$R - O - R'$$

dimethyl ether

$$CH_3 - O - CH_3$$

methyl phenyl ether

$$CH_3 - O - \bigcirc$$

structure of **epoxides**

$$\begin{array}{c} O \\ \diagup \diagdown \\ R - CH - CH - R' \end{array}$$

$$\begin{array}{c} O \\ \diagup \diagdown \\ CH_2 - CH_2 \end{array}$$

ethylene oxide

structure of **esters**

$$R - C\!\!\begin{array}{c} \diagup O \\ \diagdown OR' \end{array}$$

1,2-dihydroxybenzene (catechol)

$$\bigcirc\!\!\begin{array}{c} OH \\ OH \end{array}$$

$$C \equiv N$$

$$\bigcirc$$

nitrile benzonitrile

isonitriles (*n*) organic compounds (p.16) containing the isocyanide functional group (p.36) –NC in the molecular structure. They are prepared by reacting alkyl iodides with silver cyanide. Isonitriles are highly poisonous, colourless liquids with unpleasant odours. They are not very soluble in water, and can be reduced to produce secondary amines (p.38).

RI + AgCN → RNC + AgI
formation of **isonitriles**

cyanides (*n*) normally refers to inorganic compounds containing the cyano (p.37) group, e.g. sodium cyanide NaCN, potassium cyanide KCN. The corresponding organic compounds are called nitriles (p.40).

isocyanides (*n*) normally refers to inorganic compounds containing the isocyanide functional group –NC.

diazonium salts any salt formed between an aromatic compound (p.16) possessing an azo group (p.38) and an inorganic acid radical, an anion (p.126). Preparation of these highly reactive compounds is carried out below 0°C by treating aromatic primary amines (p.38) with nitrous acid. They are used as synthetic intermediates and in the manufacture of dyes.

amino acids a group of synthetic and naturally occurring organic compounds possessing both an amino group (p.38) and a carboxylic acid group (p.36). In the twenty-seven naturally occurring amino acids the two functional groups (p.36) are attached to the same carbon atom and the molecules can act both as amines and as acids. Amino acids are combined together in animal bodies to form proteins, one amino acid being joined to another through the peptide link, –CO–NH–, by the carboxylic acid of one amino acid being joined to the amino group of the other.

organometallic compounds an extensive group of chemical compounds (p.16) in which metal atoms are linked directly to a carbon atom in an alkyl (p.35) or aryl (p.36) group, e.g. methyl lithium $LiCH_3$, phenyl magnesium bromide C_6H_5–Mg–Br. These compounds are very useful as synthetic intermediates and are usually kept in solution in ethers (p.40). They react readily with such compounds as aldehydes (p.39) and ketones (p.39).

diazonium salts
benzenediazonium chloride

amino acids

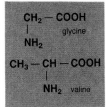

carbohydrates (*n*) a group of closely related compounds possessing a general formula (p.35) $C_n(H_2O)_m$. Many carbohydrates are considered as polyhydric alcohols (p.39) although their true structures are more complex than this. They are usually divided into two main groups, sugars and polysaccharides (↓).

aldose (*n*) a carbohydrate (↑) with a terminal aldehyde group and commonly with a general formula $C_nH_{2n}O_n$. These are sugars which are capable of reducing Fehling's solution. Aldoses are optically active and a large number of stereoisomers are frequently possible.

ketose (*n*) a carbohydrate (↑) with a ketone (p.39) carbonyl group (p.36) in the carbon chain. These are sugars which are capable of reducing Fehling's solution. Ketoses are optically active and possess several chiral centres (p.32) among the carbon atoms with a general formula (p.35) of $C_nH_{2n}O_n$.

polysaccharide (*n*) a group of long chain (p.33) carbohydrates (↑) that are non-crystalline, colloidal (p.174), insoluble in water and generally tasteless. Most polysaccharides are naturally occurring polymers (p.236), e.g. starch and cellulose, built up from ketose (p.42) units each consisting of a chain or ring of six carbon atoms. Polysaccharides have the general formula (p.35) $C_n(H_2O)_m$.

chromophore (*n*) a chemical group (p.51) in a compound which absorbs a specific wavelength of light causing the compound to reflect a characteristic colour. Chromophores, such as azo groups (p.38), are deliberately incorporated into synthetic dyes in order to create special colours. **chromophoric** (*adj*).

auxochrome (*n*) a chemical group (p.51) which does not itself absorb visible light, but serves to enhance the colour or alter the wavelength of the light absorbed by a chromophore (↑). Auxochromes tend to shift the wavelength of the absorbed light towards the red end of the spectrum so that substances look greener or bluer. Hydroxyl groups (p.36) and halogen groups (p.37) act as auxochromes.

leuco base a colourless compound produced by reducing a dye. Alkaline solutions of leuco dyes are used as the first stage in a dyeing process for dyes which are normally insoluble. Once the leuco base has become attached to the fabric it is oxidized back to the original coloured dye.

leuco compound another name for a **leuco base** (↑).

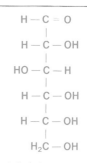

carbohydrates
glucose – an **aldose**

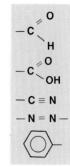

chromophore

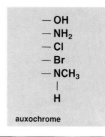

auxochrome

simple structure of an atom

electron shell

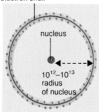

nucleus

10^{12}–10^{13}
radius
of nucleus

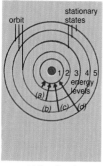

orbit

stationary
states

1 2 3 4 5
energy
levels

(a)

(b) (c) (d)

(a) $E_2 - E_1 = h\nu_a$
(b) $E_3 - E_1 = h\nu_b$
(c) $E_4 - E_1 = h\nu_c$
(d) $E_5 - E_1 = h\nu_d$

Bohr atom

particle (n) a very small piece of matter, which has a mass, but is usually considered to be too small to have dimensions. **particulate** (adj).

atomic structure an atom consists of a positively charged nucleus, surrounded by electrons. The negative charge on the electrons is equal to the positive charge on the nucleus making the atom neutral. The electrons are arranged in shells and define the volume of the atom. The radius of an atom is approximately 10^{12} to 10^{13} times greater than the radius of its nucleus. Almost all of the mass of the atom is concentrated in the nucleus.

Bohr atom an atomic structure to account for the atomic spectra (p.43) of hydrogen. Bohr suggested (a) the electrons could revolve only in selected orbits, called **stationary states** and did not radiate energy; (b) each stationary state corresponded to a certain energy level; (c) movement of an electron from one stationary state to another of lower energy produced radiation of a definite frequency, such that $E_1 - E_2 = h\nu$, where h = Planck's constant, ν = frequency of the radiation.

sub-atomic particle any one of the particles from which an atom is composed, e.g. electrons, protons, etc.

fundamental particle same as **sub-atomic particle** (↑).

fundamental (adj) describes processes, ideas or objects over which there is no choice and which form the basis for the development of other processes or ideas, e.g. fundamental particles (↑) are the basis of the structure of an atom. Contrast **basic** (↓).

basic (adj) (1) describes a fact, idea, or process that is chosen to form the basis of an argument, or theory. (2) describes a compound which neutralizes an acid.

charge (n) a fundamental property of sub-atomic particles (↑). It is of two kinds, called negative and positive. Electric charge is discrete (p.44). Like charges repel each other, unlike charges attract each other. The quantity of electric charge is measured in coulomb.

electron (n) a sub-atomic particle with a negative charge of 1.602×10^{-19} coulomb, and a **rest mass** of 9.107×10^{-28} g. The electron can be made to move by electrical attraction or repulsion, and it obeys the laws of wave motion when it moves, in accordance with de Broglie's equation. An electron can never be at rest, so it always possesses kinetic energy and can possess potential energy as well. The symbol for electron is e.

discrete (*adj*) describes matter or energy that cannot be
continuously subdivided, but has a limit of a unit that
cannot be further divided, e.g. the charge on an
electron (p.43) is the smallest charge there is, it cannot
be divided into a smaller charge; matter is composed of
atoms which cannot be divided and still recognized as
that particular type of matter.

Millikan's apparatus an apparatus to determine the
charge on an electron. Two parallel metal plates, held
accurately by glass insulators, have a potential applied
to them. Oil drops are sprayed through a hole in the
upper plate. The drops are charged by friction, and
further charged by X-rays. With the metal plates
earthed, the oil drops fall under gravity at a terminal
velocity, measured by observations through a
microscope. The charged metal plates hold an oil drop
stationary. From the upward electrical force on a
stationary drop and the gravitational force on the drop,
the charge on an oil drop can be calculated. All such
charges are found to be an integral multiple of 1.602×10^{-19} coulomb, which is taken to be the charge on a
single electron.

electron energy levels the energy level of an electron
can have specific values which vary from 0, when it is
just free of the nucleus of an atom, to a value for its
lowest energy state, where it is at its nearest to the
nucleus. The electron can exist only in these definite
energy levels.

positron (*n*) a fundamental particle (p.43) with the same
mass, and a positive charge of the same magnitude, as
an electron. It is represented by e^+. Free positrons are
stable, but take part in pair-annihilation.

proton (*n*) a sub-atomic particle (p.43) with a positive
charge equal in magnitude to that of an electron. Its
mass is 1.672×10^{-24}g, i.e. approximately 1840 times
the mass of an electron. Protons are present in all
atomic nuclei. A proton is represented by the symbol p.

neutron (*n*) a sub-atomic particle (p.43) having no
charge, and a mass of 1.675×10^{-24}g, that is
approximately the same as that of a proton ($\uparrow$).
Neutrons are present in all atomic nuclei except those
of hydrogen. Neutrons are stable in nuclei, but outside
a nucleus a neutron decays to give a proton and an
electron; the half-life (p.183) of a free neutron is
approximately 780 seconds. A neutron is represented
by the symbol n.

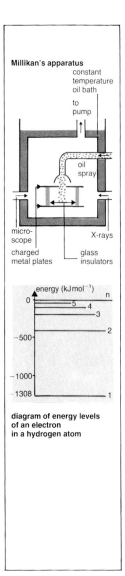

Millikan's apparatus

diagram of energy levels
of an electron
in a hydrogen atom

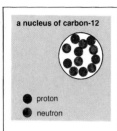

a nucleus of carbon-12

● proton
● neutron

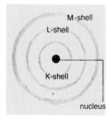

electron shells

M-shell
L-shell
K-shell
nucleus

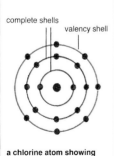

complete shells
valency shell

a chlorine atom showing
the valency shell and
the complete shells

nucleus (*n. pl. nuclei*) a nucleus, except that of hydrogen, is made up of protons (↑) and neutrons (↑). Nuclei in their ground state are almost spherical, and have a radius, r, which depends on a constant, r_o, and A, the mass number (p.47), thus: $r = r_o A^{1/3}$. The value of r_o is approximately 10^{-15} m. **nuclear** (*adj*).

atomic mass unit one-twelfth of the mass of the isotope $^{12}_{6}C$, which is used in defining relative atomic mass (p.29); it is approximately 1.6605×10^{-27} kg.

a.m.u. (abbr) atomic mass unit (↑). A proton and a neutron have a mass of approximately 1 a.m.u.; an electron is approximately 1/1840 a.m.u.

fermi (*n*) a unit of length, 10^{-15} m, approximately the value of r_o in measurements of nuclear radii. All nuclei have a radius of the order of a few fermi. *See* **nucleus** (↑).

extra-nuclear structure the structure of an atom outside its nucleus. It describes the arrangement of electrons in electron shells (↓) surrounding a nucleus.

electron shell a space in which electrons are found, all of which have the same principal quantum number (p.86). The electrons occupy energy sub-levels within the main energy level of a shell. The shells are designated by letters: K, L, M, etc. (↓); they form the extra-nuclear structure (↑).

quantum shell another name for **electron shell** (↑).

K-shell (*n*) the innermost electron shell (↑) with a principal quantum number of 1. It contains a maximum of 2 electrons. It is called a K-shell because the K-lines of the X-ray spectrum (p.92) of an element result from electron transitions from outer shells to the K-shell.

L-shell (*n*) the electron shell (↑) next to the K-shell (↑); it has a principal quantum number of 2 and contains a maximum of 8 electrons. The L-lines of the X-ray spectrum of an element result from transitions of electrons from outer shells back to the L-shell.

M-shell (*n*) the electron shell (↑) next to the L-shell (↑); it has a principal quantum number of 3, and contains a maximum of 18 electrons. It gives rise to the M-lines of an X-ray spectrum.

valency shell the outermost shell of an atom. It can be a K, L, M, or N, etc. electron shell (↑). The electrons in the shell take part in chemical bonds (p.69).

complete shell an electron shell (↑) which contains its maximum number of electrons. A complete shell cannot take part in chemical bonds (p.69). It is an inner shell to a valency shell (↑).

screening (n) the positive charge, Z, on a nucleus
provides an electrostatic attraction for electrons in
electron shells (p.45). Complete shells (p.45) screen
the nucleus from extra-nuclear electrons. Same-shell
electrons also contribute to this screening effect. The
effective nuclear charge, Z_{eff}, is less than Z, as
$Z_{eff} = Z - S$, where S is a screening constant,
dependent on the number of electrons present, and the
orbitals they occupy.

shielding (n) same as **screening** (↑).

mass spectrograph ions are produced by bombarding
the vapour of an element with electrons. The positive
ions pass through slits S_1 and S_2, and are acted upon
by opposing electric and magnetic fields, so that all
ions emerge from slit S_2 with the same velocity, v. The
positive ions enter an evacuated semicircular chamber,
and are acted upon by a second magnetic field,
causing them to follow a semicircular path, of radius r. If
E is the charge on the ion, H the magnetic field strength,
and m the mass of the ion, then $m/E = Hr/v$. If the
charge on all positive ions from the element is the same,
then $m \propto r$, as E, H, and v are all constant. The ions
produce a line on a photographic plate, and the
thickness of the line is proportional to the proportion of
the isotope (↓) of the element. The value of r is
calibrated against known standards, e.g. $^{12}_{6}C$. This
procedure is called **velocity focusing**; it determines
the relative mass of the isotopes of an element and an
approximate value of their proportion. This is produced
by a **mass spectrometer**.

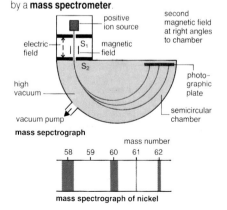

mass sepctrograph

mass spectrograph of nickel

mass number the nearest whole number to the mass of an isotope (↓), *see* **mass defect** (p.102). The mass number is equal to the number of protons and neutrons in a nucleus. Mass number is represented by the symbol A, so $A = p + n$ for a nucleus. e.g. deuterium is an isotope of hydrogen, $A = 2$ as the nucleus consists of 1 proton and 1 neutron.

atomic number a number equal to the number of protons in a nucleus of an atom of an element; it has the symbol Z. It is equal to the magnitude of the positive charge on the nucleus. *See* **screening** (↑).

isotopes (*n*) nuclear species which have the same atomic number (↑) but different mass numbers (↑), hence have the same number of protons, but different numbers of neutrons. Many elements have isotopes, i.e. atoms with the same properties but different relative atomic masses. The presence of isotopes explains the divergence of relative atomic masses from whole numbers. Isotopic forms of the same element have the same number of protons, hence they must have the same number of electrons in the same extra-nuclear structure (p.45). **isotopic** (*adj*).

isotopes

element	mass number	number of		isotopic ratio (%)
		protons	neutrons	
carbon	12	6	6	98.9
	13	6	7	1.1
magnesium	24	12	12	78.6
	25	12	13	10.1
	26	12	14	11.3
chlorine	35	17	18	75.4
	37	17	20	24.6
copper	63	29	34	69.0
	65	29	36	31.0

relative abundances of isotopes of nickel

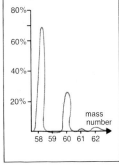

isotopic ratio the ratio of the different isotopes of an element, usually expressed as a percentage. The isotopic ratio for elements obtained from natural sources is always the same, e.g. isotopes of nickel are:

^{58}Ni —67.9% ^{60}Ni —26.2% ^{61}Ni —1.2%
^{62}Ni —3.7% ^{64}Ni —3.7%.

The relative atomic mass of nickel is the weighted average of these isotopes.

nuclear charge the charge on a nucleus is Ze, where e is the charge on a proton equal in magnitude, but opposite in sign, to that on an electron, and Z is the atomic number (p.47) of an element. The nuclear charge exerts an attractive force on the extra-nuclear electrons, and the total negative charge on all electrons is equal in magnitude to the nuclear charge.

Davisson and Germer's experiment an electron gun produces a beam of electrons, all with the same velocity, brought about by an accelerating potential. The beam strikes a crystal of nickel, is diffracted and detected by a Faraday cylinder connected to a galvanometer. For a given angle of incidence, the intensity of the beam of diffracted electrons varies with the accelerating potential as shown in the graph. The energy of an electron $= Ve = \frac{1}{2}mv^2$, where V is the accelerating potential, e the charge on an electron, m the mass, and v the velocity of an electron. The momentum of an electron is mv, and $mv \propto \sqrt{V}$. Now $mv = h/\lambda$ by de Broglie's equation (p.85). Therefore $1/\lambda \propto \sqrt{V}$. The graph of the intensity of a beam in X-ray diffraction by a crystal corresponds to the graph for electron intensity, so the experiment demonstrates that a beam of electrons is diffracted by a crystal.

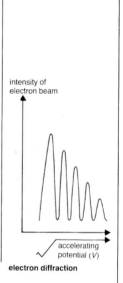

intensity of
electron beam

accelerating
potential (V)

electron diffraction

Davisson and Germer's experiment

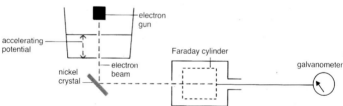

electron
gun

accelerating
potential

nickel
crystal

electron
beam

Faraday cylinder

galvanometer

Schrödinger's equation the equation for an electron in a one-electron atom. The solutions to the equation have significance only for certain values of E (The total energy for each atomic energy level). Schrödinger's equation is also called a **wave equation**.

wave function a solution to Schrödinger's equation. The solution gives the probability of finding an electron of energy E in a specified volume. It is a mathematical description of an orbital (↓).

eigenvalue (*n*) the value of *E* acceptable in Schrödinger's equation (↑).

eigenfunction (*n*) the wave function (↑) for an eigenvalue.

probability distribution the idea of an electron as a particle is replaced by the idea of a charge cloud of varying charge density with a wave pattern round a nucleus. If ψ is a wave function (↑), then ψ^2 is the density of electron per unit volume and $-e\psi^2 dv$ is the effective electron charge associated with a volume *dv*.

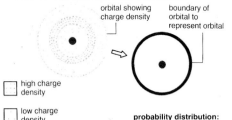

probability distribution: charge density of an orbital

orbital (*n*) the region of space occupied by a charge cloud of an electron, with a probability distribution (↑). Different orbitals have different shapes, and each orbital can be occupied by 2 electrons, with opposing spin states at one and the same time.

s-orbital (*n*) an orbital with radial symmetry. The charge cloud of electrons is not concentrated in any direction. It has the lowest energy level in an electron shell (p.45).

p-orbital (*n*) the p-orbitals have a dumb-bell shape, and a directional character along each of the three axes in space. Each orbital has a nodal plane which has a zero probability of finding an electron. There are three p-orbitals in each shell.

d-orbital (*n*) there are five d-orbitals, each possessing a directional character. Three are similar, i.e. d_{xy}, d_{yz}, d_{xz}, and the other two differ.

f-orbital (*n*) there are seven f-orbitals, three dissimilar and four similar in shape.

s-electron (*n*) an electron in an s-orbital. There can be two s-electrons each with opposing spin.

p-electron (*n*) an electron in a p-orbital. There can be two p-electrons in each p-orbital; they have opposing spin. In accordance with Hund's rules, the first three electrons enter each of the three p-orbitals; these electrons have parallel spin. The next three fill the p-orbitals.

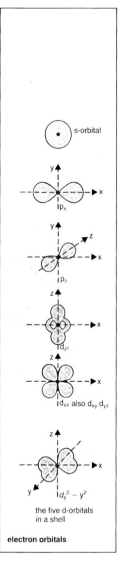

the five d-orbitals in a shell

electron orbitals

d-electron (*n*) an electron in a d-orbital. There can be two d-electrons in each d-orbital, they have opposing spin. In accordance with Hund's rules, the first five d-electrons enter each of the five d-orbitals, these electrons have parallel spin. This half-filled d-orbital level is relatively stable. Some d-electrons take part in electrovalent bonds and are responsible for variable electrovalency in transition elements (p.52).

orbital hybridization s- and p-orbitals can combine together to form **hybrid orbitals**, described from the number of s- and p-orbitals taking part in orbital hybridization. One s- and three p-orbitals form four sp^3 orbitals, each of which are identical in energy level, when an atom is in an excited state, e.g. a carbon atom in its ground state ($\downarrow$) has 2 s-electrons and 2 p-electrons in its valency shell; these form four sp^3 orbitals in an excited state. As the orbitals are identical, they are equally inclined in space, at 109.5° to each other, in agreement with the tetrahedral bond directions of carbon. Other hybridizations include sp^2 hybrids, three coplanar hybrids inclined at 120° to each other; sp hybrids, two collinear hybrid orbitals; dsp^2 hybrids, with four hybrid orbitals which are square coplanar; d^2sp^3 hybrids which have an octahedral arrangement.

hybrid orbital *see* **orbital hybridization** (↑).

ground state an atom, or molecule, is in its ground state when every electron is in its state of least energy. On excitation, electrons move into states of higher energy. On falling back to a ground state, an electron emits radiant energy forming a spectral line.

Sidgewick-Powell approach a method of determining the configurations of atoms round a central atom; it has four basic assumptions: (1) a bond between two atoms holds two electrons. (2) electron pairs, whether in bonds, or in lone pairs, repel each other. (3) the repulsive force from a lone pair is greater than that from a bonding pair. (4) in double bonds, four electrons act as a single bonding pair. Application of these facts determines the configuration of a molecule.

periodic table an arrangement of elements in order of ascending atomic number such that elements with like properties are placed in columns or rows.

periodic law also called the law of periodicity. The periodic recurrance of similar chemical properties when elements are arranged in order of ascending atomic number.

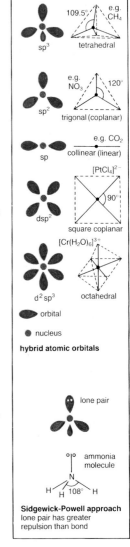

sp^3 — 109.5° — e.g. CH_4 — tetrahedral

sp^2 — e.g. NO_3 — 120° — trigonal (coplanar)

sp — e.g. CO_2 — collinear (linear)

dsp^2 — $[PtCl_4]^{2-}$ — 90° — square coplanar

d^2sp^3 — $[Cr(H_2O)_6]^{3+}$ — octahedral

⬬ orbital

● nucleus

hybrid atomic orbitals

lone pair

ammonia molecule

H — N — H — 108° — H

Sidgewick-Powell approach
lone pair has greater repulsion than bond

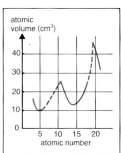

atomic volume of elements
(excluding gases)

periodic system the system of classification of elements based on the periodic table (↑). The atomic number (p.47) of an element is equal to the number of extra-nuclear electrons, so the system is really based on extra-nuclear structure (p.45).

periodicity (*n*) the regular occurance of similar chemical and physical properties with increasing atomic number (p.47) of elements, e.g. atomic volumes (p.53), electrovalency and covalency. *See* graph for atomic volume.

group (*n*) a column in the periodic table (↑) containing elements with similar properties, e.g. the halogens in group VII, consisting of fluorine, chlorine, bromine, and iodine.

period (*n*) a row of elements in the periodic table (↑). There is a gradual change of chemical nature from characteristic metals on the left to characteristic non-metals on the right. Electrovalency increases from +1 to +3 from left to middle and covalency decreases from 4 to 1 from middle to left, of a period. Electrovalencies of the same magnitude as covalencies decrease from −2 to −1. The last element in a period is a noble gas with zero valency.

a period in the periodic table

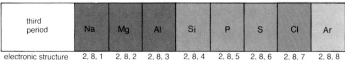

third period	Na	Mg	Al	Si	P	S	Cl	Ar
electronic structure	2, 8, 1	2, 8, 2	2, 8, 3	2, 8, 4	2, 8, 5	2, 8, 6	2, 8, 7	2, 8, 8

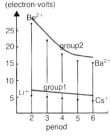

ionization energies of ions

ionization energy the energy required to remove an electron from a gaseous atom leaving a positive ion. The ionization energies for s- and p-orbitals in the L-shell of atoms is shown in the graph; it is the energy required to remove one electron from an atom. The increasing ionization energy from 2s to $2p^6$ indicates the reason for the change from positive electrovalency to covalency in a period (↑). The graph showing the ionization energies of metals in groups 1 and 2 indicates (1) the increase in electropositive properties as a group is descended. (2) the stronger electropositive nature of metals in group 1 compared with those of group 2. (3) the shielding effects of inner-shell electrons. The increasing ionization energies required to remove electrons from an atom indicate the increasing attraction of the nucleus for the remaining electrons as successive electrons are removed.

ionization potential an alternative term for **ionization energy** (p.51).

atomic volume relative atomic mass divided by density. Atomic volume shows a periodic relation with atomic number.

s-block element an element whose valency shell contains s-electrons (p.49) only. Such elements are found in groups 1 and 2 of the periodic table (p.50) and they readily form cations (p.126).

p-block element an element whose valency shell contains both s- and p-electrons. Such elements are found in groups 3 to 7 of the periodic table (p.50); they form covalent bonds, and in groups 6 and 7 they form anions (p.126).

manganese - transition element

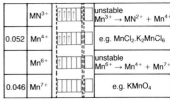

atom/ion		orbitals		remarks
(nm) radius	symbol	3d	4s	
0.137	Mn	↑↑↑↑↑	↑↓	element
0.080	Mn²⁺	↑↑↑↑↑		e.g. MnSO₄

	atom/ion	orbitals	remarks
	MN³⁺		unstable Mn³⁺ → MN²⁺ + Mn⁴⁺
0.052	Mn⁴⁺		e.g. MnCl₂.K₂MnCl₆
	Mn⁶⁺		unstable Mn⁶⁺ → Mn⁴⁺ + Mn⁷⁺
0.046	Mn⁷⁺		e.g. KMnO₄

☐ empty orbital   ↑ ↑ parallel spin   ↑ ↓ opposite spin   electron pairs

d-block element an element whose valency electrons are s-electrons in the outermost shell, and d-electrons in the penultimate shell.

transition element a d-block element (↑) placed between the s- and p-block elements. It exhibits variable electrovalency arising from the possibility of both s- and d-electrons (p.49) being removed to form cations. The possible electrovalent states for manganese are indicated in the diagram.

lanthanide (*n*) one of 14 elements with almost identical properties, which follow lanthanium in the periodic table. They are formed by filling the 5 f-orbitals of an atom.

actinide (*n*) one of 14 elements with almost identical properties which follow actinium in the periodic table. They are formed by filling the 6 f-orbitals.

lanthanide contraction the atomic radius of elements after the lanthanides (↑) is smaller than expected because of the greater decrease in atomic radius as the 5 f-orbitals of the lanthanides are filled; this is the lanthanide contraction.

radiation (*n*) the transmission of energy by electromagnetic waves or by a stream of particles that have a wave-like nature. **radiate** (*v*), **radiant** (*adj*).

ray (*n*) the straight line path followed by radiation from a source. It is often used to replace radiation, but the use is not entirely correct.

radioactivity (*n*) the nuclei of certain elements disintegrate spontaneously and emit ionizing radiation which can be one or two of three types, namely, alpha, beta and gamma radiation. The process is called radioactivity. **radioactive** (*adj*).

radiation	nature	electrical charge	speed (as % speed of light)	relative ionizing power	relative penetration
α-rays	helium nuclei	+2 units	.5	10 000	1
β-rays	electrons	−1 unit	3.99	1000	100
γ-rays	electromagnetic waves	no charge	100	1	10 000

types of radiation from radioactivity

alpha radiation a stream of alpha particles emitted by a radioactive element. The emission can be accompanied by gamma radiation but not by beta radiation (↓). The radiation carries a positive charge.

beta radiation a stream of beta particles emitted by a radioactive element. The emission can be accompanied by gamma radiation (p.54) but not by alpha radiation (↑). The radiation carries a negative charge.

α**-rays** (*n*) *see* **alpha radiation** (↑).

β**-rays** (*n*) *see* **beta radiation** (↑).

γ**-rays** (*n*) *see* **gamma radiation** (p.54).

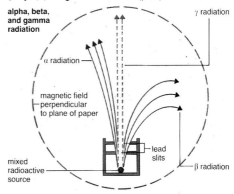

alpha, beta, and gamma radiation

gamma radiation electromagnetic radiation of high
frequency, emitted as a result of the nucleons (p102) in
certain radioactive nuclei undergoing reorganization.
Gamma radiation has a high penetrating power; it will
penetrate approximately 150mm of lead. It has a low
ionizing power, and has no electrical charge, so it is not
deflected by electric or magnetic fields. Like other e.m.
radiations, it can be considered to consist of photons
(p.85) travelling at the speed of light. The frequencies of
gamma radiation are related to nuclear energy levels
and form a nuclear spectrum characteristic of the
emitting nucleus. Gamma radiation can accompany
either alpha or beta radiation.

alpha particle a helium nucleus of 4 a.m.u. (p.45) and
electrical charge +2 units. The initial speed of emission
is of the order of $1.6 \times 10^7 \, \text{ms}^{-1}$ and the particles have
a range ($\downarrow$) in air of about 2 to 10cm. Alpha particles cause intense ionization in gases through which they
pass; they can be detected by scintillations on a
fluorescent screen, by a photographic plate, or by a
Geiger counter (p.57).

beta particle a high-energy electron with a speed from
3% up to 99% that of light. An electron with a high
speed has a relatavistic mass. The particles have an
average range in air of approximately 7.5m. Beta
particles produce weak ionization in gases through
which they pass; they can be detected by a
photographic plate, or by a Geiger counter (p.57).

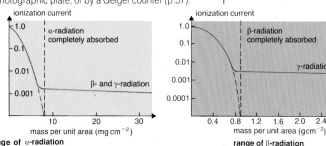

range of α-radiation

range of β-radiation

range (*n*) the maximum distance anything can travel.
The range of radioactive radiations depends on the
mass per unit area of the material absorbing them. For
α-radiation, the range is $8.0 \, \text{mg cm}^{-2}$; for β-radiation it is
$0.9 \, \text{g cm}^{-2}$; for γ-radiation, the range is theoretically
infinite as absorption is never complete.

no gamma radiation

$$^{210}_{84}Po \rightarrow\ ^{206}_{82}Pb\ +\ ^{4}_{2}He^{2+}$$

gamma radiation

$$^{238}_{92}U \rightarrow\ ^{234}_{90}Th + \alpha + \gamma$$

alpha emission

no gamma radiation

$$^{215}_{83}Bi \rightarrow\ ^{215}_{84}Po\ +\ e\ +\ \bar{\gamma}$$

gamma radiation

$$^{214}_{82}Pb \rightarrow\ ^{214}_{83}Bi + e + \gamma^{-} + \gamma$$

beta emission (electrons)

ion-pair[1] (n) an ionizing radiation passing through a gas removes extra-nuclear electrons from atoms or molecules leaving positive gaseous ions. The separate positive ions and electrons are ion-pairs.

specific ionization the number of ion-pairs (↑) produced per mm of track by an ionizing radiation passing through air at 15°C and 1 atmosphere pressure. It is a measure of the ionizing capability of the radiation.

alpha emission a process in which a radioactive nucleus emits an alpha particle, and which may or may not be accompanied by the emission of gamma radiation (↑). The resulting nucleus has a mass number 4 a.m.u. less, and an atomic number 2 less; this corresponds to a loss of two protons and two neutrons. Alpha emission is mainly confined to the heavier nuclei.

	neutron ➡	proton	+ electron +	anti-neutrino
mass	n amu	n amu	0	0
charge	0 ➡	+1	−1	0
spin	± ½	± ½	± ½	± ½

beta emission   (β− emission)

beta emission a process in which a radioactive nucleus emits a beta particle (↑) and which may or may not be accompanied by the emission of gamma radiation (↑). The emission is accompanied by the conversion of a neutron in the nucleus to a proton and an electron. The relationship of mass, charge and spin is shown in the diagram, and postulates an anti-neutrino being emitted. The resulting nucleus, after emission, has no change in mass number but the atomic number *increases* by one. Beta emission includes the emission of a positron (e^{+}) in which a proton in the nucleus is converted to a neutron emitting a positron and a neutrino. The resulting nucleus has no change in mass number, but the atomic number *decreases* by one. Nuclei, both heavy and light, can produce beta emission, but emission of a positron is limited to artificial radioactive elements. Positron emission is called β^{+} emission.

gamma emission after a nuclear process, such as alpha or beta emission, has occurred, the resulting nucleus may be excited. The excess energy is released as gamma radiation (↑) and the nucleus returns to its ground state (p.50).

emit (v) to give out, without an agent taking part in the process, energy, radiation, sound, gas or odour, from a source. **emission** (n), **emitter** (n).

bombard (v) to hit a relatively large object many times with many smaller objects. The bombarding objects possess a high energy, e.g. a stream of electrons bombarding a metal target. **bombardment** (n).

deflection (n) a change in direction of a moving object, beam of radiation, or stream of particles, caused by an agent. Attention is drawn to the change in direction, e.g. alpha radiation is deflected by an electric field.

divergence (n) the continuous separation of two or more paths. Attention is drawn to the gradual separation and the continuity of the paths, e.g. the divergence of alpha and gamma radiation under the influence of a strong magnetic field. **diverge** (v), **divergent** (adj).

scattering (n) the process of separating individual units of a whole, or of a group, from one another and spreading them over an area, or a volume, in a random manner. Attention is drawn to the process and the random nature of the result, e.g. an exploding gas cylinder results in the scattering of pieces of cylinder over a wide area. **scatter** (v).

track (n) the continuous visible signs left along the path of an object, particle or radiation.

cloud chamber a chamber, with a plastic or glass window, filled with a gas saturated with water vapour. The chamber is closed by a piston. A valve separates the piston from an evacuated bulb. When the valve is opened, the piston drops, increasing the volume of the chamber. This cools the gas which becomes super-saturated with water vapour. The valve also operates a camera, which photographs the tracks (↑) made by ionized particles. The radiation from the radioactive source ionizes molecules of the gas and the water vapour condenses in tiny droplets on the ionized particles, forming tracks. The tracks of alpha, beta and gamma radiation can be studied using the apparatus and also the interaction of the radiation with molecules of different gases.

bubble chamber a modern development of the cloud chamber (↑) in which radiation passes through a superheated liquid, e.g. liquid hydrogen or liquid methyl butane. This produces a track consisting of a series of bubbles, and provides a better visual track than a cloud chamber.

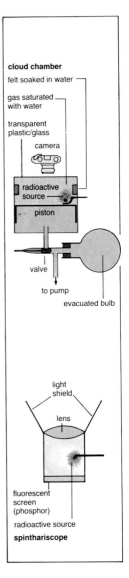

cloud chamber

felt soaked in water

gas saturated with water

transparent plastic/glass

camera

radioactive source

piston

valve

to pump

evacuated bulb

light shield

lens

fluorescent screen (phosphor)

radioactive source

spinthariscope

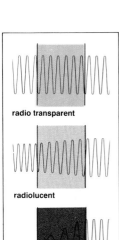

radio transparent

radiolucent

radio opaque

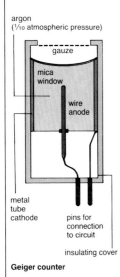

argon
(¹/₁₀ atmospheric pressure)

gauze

mica
window

wire
anode

metal
tube
cathode

pins for
connection
to circuit

insulating cover

Geiger counter

scintillation (n) a flash of light produced when an electromagnetic radiation or an alpha particle hits a phosphor (↓). **scintillate** (v).

phosphor (n) a substance which absorbs electromagnetic radiation, or a form of particle energy, and emits immediately an electromagnetic radiation in the visible spectrum, e.g. zinc sulphide. Such a substance is fluorescent. Phosphors are used in television tubes.

spinthariscope (n) a device for counting the number of alpha particles emitted by a radioactive source by counting the number of scintillations (↑) the particles make on a fluorescent screen coated with phosphor (↑). An approximate measurement is obtained.

scintillation counter phosphors (↑) transparent to the light they emit, and in the form of liquids and even gases, in addition to solids, are used in scintillation counters. The feeble scintillations are converted by photomultipliers into electrical pulses and counted. These devices are sensitive and give fast, accurate counting of particles for all types of radiation.

radio transparent describes a material that allows electromagnetic radiation to pass through with little or no reduction in intensity.

radiolucent (adj) describes a material that allows electromagnetic radiation to pass through with some reduction in intensity, e.g. flesh is radiolucent to X-rays, while bone is radio opaque (↓).

radio opaque describes a material that absorbs most or all of an electromagnetic radiation. Materials containing atoms of high relative atomic mass are strong absorbers. Radiation of high frequency has a greater penetrating power than that of low frequency. Radio opaqueness also depends on the width of the material.

Geiger counter a device for detecting radiation. A thin wire acts as an anode inside a cylindrical metal cathode. The electrodes are enclosed in an insulating cover containing argon at 1/10 atmospheric pressure. A p.d. of 1000 to 1500 V is applied to the electrodes. Ionizing radiation enters through the mica window, ionizes the gas, producing ion-pairs. The ions, accelerated by the electrical field, produce secondary ionization so that one ionizing particle can produce up to 10^8 ion-pairs. The ions are discharged producing a current pulse. The pulses are counted electrically. The counter does not distinguish between α, β or γ radiation, all produce the same result for the same intensity.

Geiger-Müller tube *See* **Geiger counter** (p.57).

penetrate (*v*) to go into, or through, a material because of energy being used in the process, or because of a force being applied. Attention is drawn to that which penetrates, e.g. beta radiation penetrates further than alpha radiation. **penetration** (*n*).

biological shield a wall of concrete, several metres thick, lined with steel, to prevent radiation escaping from a nuclear reactor (p.66). The shield is radio opaque (p.57).

radioactive disintegration a nucleus of a radioactive species undergoes a spontaneous change forming a nucleus of a new element. The process is accompanied by the emission of an alpha or a beta particle, and may be accompanied by gamma radiation. The new element may be stable or undergo radioactive disintegration. A beta particle is either an electron or a positron. Disintegration is a random process.

daughter element the new element produced by radioactive disintegration (↑).

rate of disintegration the natural disintegration of a radioactive element is unaffected by any chemical agent or physical condition. The rate is directly proportional to the number of atoms present, and cannot be speeded up or retarded. The rate of

1 g radium emits 3.608×10^{10} α-particles per sec

relative atomic mass Ra = 226.0

no. of atoms in 1 g Ra

$$= \frac{1}{226.0} \times 6.02 \times 10^{23} = N$$

decay rate $= -\frac{dN}{dt} = 3.608 \times 10^{10} \ s^{-1}$

$$-\frac{dN}{dt} = \lambda N$$

$$\lambda = \frac{3.608 \times 10^{10} \times 226.0}{6.02 \times 10^{23}}$$

$$= 1.354 \times 10^{-11} \ s^{-1}$$

disintegration for N atoms of an element is $-dN/dt$ and
this is proportional to N, so $-dn/dt = \lambda N$
On integration, $\ln N = -\lambda t + \ln N_0$, where λ is the decay
constant. Taking common logarithms and substituting
for the half-life period (p.60) by $\lambda = 0.301/t_{1/2}$.
$\log N = \log N_0 - 0.301 t/t_{1/2}$, where N_0 is the initial
number of atoms.

radioactive decay radioactive elements emit radiation
and change into new elements accompanied by a
gradual decrease in their radioactivity; this is
radioactive decay. Radioactive decay is brought about
by radioactive disintegration (↑).

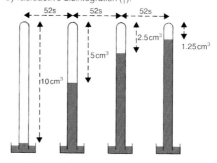

radioactive decay of 220 Rn $t_{1/2} = 52s$

decay rate the rate of disintegration (↑), the rate of
radioactive decay (↑).

activity[1] (n) the rate of disintegration (↑) of a radioactive
substance, a measure of the intensity of radioactivity.
With an initial activity A_0 the activity A after a time t is
given by: $A = A_0 e^{-\lambda t}$ where λ is the decay constant
(p.60). If radioactive decay (↑) produces a radioactive
daughter element (↑) with a greater decay constant
than that of its parent, where A_1 and λ_1 refer to the
parent and A_2 and λ_2 refer to the daughter, then

$$\frac{A_2}{A_1} = \frac{\lambda_2}{\lambda - \lambda_1}.$$

exponential decay from the rate of disintegration (↑), the
number of atoms of a radioactive element, N, remaining
after a time, t, from an initial number, N_0, is given by:
$N = N_0 e^{-\lambda t}$, where λ is the decay constant (p.60). This
exponential expression shows radioactivity is an
exponential form of decay, i.e. the rate of decay is
proportional to the number of remaining atoms.

decay constant a characteristic property of a radioactive species which determines the rate of disintegration (p.59) according to the equation: $-dN/dt = \lambda t$, where λ is the decay constant.

radioactive decay constant decay constant (↑), λ.

disintegration constant decay constant (↑), λ.

half-life period a characteristic constant of a radioactive species, dependent on the decay constant (↑). It is the time required for the activity (p.59) of a radioactive species to be reduced to half its value, or for the number of atoms to be reduced by half. The condition for the period of half-life ($t_{1/2}$) is $N/N_o = 0.5$, hence $e^{-\lambda t}$ $= 0.5$, i.e. $\lambda t = ln2 = 0.693$. The half-life is given by: $t_{1/2} = 0.693/\lambda$, where λ is the decay constant (↑). The half-life principle is illustrated in the diagrams; after 10 half-life periods, the remaining fraction is less than 0.1%, and decay is usually considered to be complete.

curie (n) the unit of rate of disintegration (p.59), symbol Ci, defined as a rate of 3.7×10^{10} disintegrations per second. One curie is also the amount of radioactive substance that undergoes 3.7×10^{10} disintegrations per second.

röntgen (n) the unit of quantity of radiation, symbol R. One röntgen is the quantity of gamma radiation, or X-rays, that liberates by ionization $8.38 \times 10^{-3} \text{J kg}^{-1}$ of dry air at s.t.p. A person should not be exposed to more than a total of 50mR in one year. The rate of dosage should not exceed 0.25mR per hour.

rem (abbr) röntgen equivalent man. One rem is the equivalent dose of any ionizing radiation with the same biological effect as one röntgen of X-rays.

rad (n) the unit of absorbed radiation. One rad is an energy absorption of 10^{-2}J kg^{-1} of irradiated material.

natural radioactivity the radioactivity of a member of the radioactive series (↓). The basic requirement for a spontaneous natural radioactivity is that the mass of the parent nucleus must be greater than the sum of the masses of the daughter nuclei and of any emitted particle. The loss in mass appears as an equivalent quantity of energy carried by an emitted particle or a gamma photon. Natural radioactivity decays by alpha emission, beta emission, or gamma radiation.

radioactive series there are three naturally occurring radioactive series, the thorium, uranium and actinium series. A fourth, the neptunium series, has now died out on the earth, but has been made artificially. Alpha

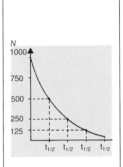

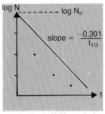

graphs for half-life period

radium $\lambda = 1.354 \times 10^{-11} \text{S}^{-1}$

$$t_{1/2} = \frac{0.693}{\lambda}$$

$$= \frac{0.693 \times 10^{11}}{1.354 \times 365 \times 24 \times 60 \times 60}$$

$$= 1620 \text{ years}$$

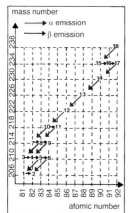

mass number

→ α emission

→ β emission

atomic number

uranium radioactive series

1 – Tl	**7** – Pb	**13** – Ra
2 – Pb	**8** – Bi	**14** – Th
3 – Tl	**9** – Po	**15** – Th
4 – Pb	**10** – Po	**16** – Pa
5 – Bi	**11** – At	**17** – U
6 – Po	**12** – Rn	**18** – U

emission results in a decrease of 4 a.m.u. and beta emission has only a very small mass change, hence all the members of each series can be represented by mass numbers of $4n$, $4n + 2$ and $4n + 3$ where n is an integer. The $(4n + 1)$ series is the neptunium series.

disintegration series *see* **radioactive series** (↑).

uranium series the $(4n + 2)$ series of the radioactive series (↑).

thorium series the $4n$ series of the radioactive series (↑). It starts with thorium–232 and ends with lead–208.

actinium series the $(4n + 3)$ series of the radioactive series (↑). It ends with lead–207.

branched disintegration a disintegration of a radioactive substance in which some nuclei emit alpha particles and some emit beta particles, e.g. $^{214}_{83}Bi$ emits an α-particle to form $^{210}_{81}Tl$ and a β-particle to form $^{214}_{84}Po$ in the uranium series (↑).

end product each radioactive series finally forms a stable non-radioactive element, which is the end product, in each case an isotope of lead.

radioactive equilibrium if a radioactive element produces a radioactive daughter element (p.58), a state of radioactive equilibrium is reached when the daughter element decays at the same rate at which it is being formed. The condition for radioactive equilibrium is: $\lambda_1 N_1 = \lambda_2 N_2$, where λ_1 and λ_2 are the decay constants (↑) for parent and daughter respectively, and N_1 and N_2 are the number of atoms of parent and daughter. For a radioactive series $\lambda_1 N_1 = \lambda_2 N_2 = \lambda_3 N_3$ ... = constant, where 1, 2, 3, etc. refer to successive generations. *See* **activity**[2] (p.121).

group displacement law when an alpha particle is emitted in a radioactive change, the product is displaced two places, or two groups, to the left in the periodic table; when a negative beta particle is emitted the product is displaced one place or one group to the right; when a positive beta particle is emitted, or a K-electron is captured, the product is displaced one group to the left. *see* **uranium radioactive series**.

Fajans and Soddy's law *see* **group displacement law**.

artificial radioactivity bombardment of nuclei by charged particles and by neutrons forms radioactive isotopes of non-radioactive elements. These isotopes have artificial radioactivity. They decay by alpha emission, positive and negative beta emission (p.55), gamma radiation (p.54), and by electron capture (p.62).

electron capture a process in which a nucleus absorbs an extranuclear electron, usually an electron from a K-shell (p.45). The nucleus is excited by particle bombardment for the process. The electron combines with a proton to form a neutron and a neutrino. The mass number of the nucleus remains the same but the atomic number decreases by 1.

induced radioactivity *see* **artificial radioactivity** (p.61).

accelerator[1] (*n*) a device which accelerates positively charged particles to high velocities and gives them high energies.

linear accelerator a series of tubes of increasing length arranged in a straight line. Alternate tubes are connected to the terminals of a high-frequency potential. The polarities of the tubes are reversed as positive particles, or ions, pass across the gaps between the tubes. A particle is then repelled by the rear tube and attracted by the front tube. The particles acquire additional energy at each gap, and are accelerated. The tube lengths and the alternating frequency are adjusted so as to accelerate the particles.

nuclide (*n*) an atomic species, defined by the number of protons and neutrons in its nucleus, by its half-life (p.183) period and type of radioactive decay (p.59).

nucleide (*n*) *see* **nuclide** (↑).

nuclear isomers nuclei with the same mass number and atomic number, but with different energy states, e.g. $^{234}_{91}$Pa has two nuclear isomers, one has a half-life of 1.17 m, the other a half-life of 6.7 h. Both emit negative beta particles, one with 2.31 MeV and the other with 1.2 MeV maximum energy.

radioactive isotopes those isotopes (p.47) of an element which are radioactive. Natural radioactivity produces radioactive isotopes, and they may be formed in different radioactive series, e.g. there are seven isotopes of polonium, atomic number 84, among which 3 come from the uranium series, 2 from the thorium series and 2 from the actinium series. Artificial radioactive isotopes are formed in nuclear reactions.

radioactive tracers a radioactive isotope (↑), usually an artificial one, used in chemical systems to allow the tracking of an element by detection of the radioactivity, e.g. $^{32}_{15}$P has a half-life of 14.3 days and is used to trace the distribution and path of phosphorus in the metabolism of living organisms.

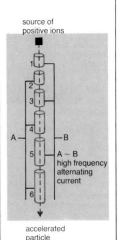

source of positive ions

A — B

A ~ B high frequency alternating current

accelerated particle

linear accelerator

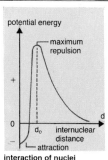

interaction of nuclei
(both positive)

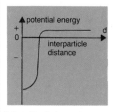

interaction between a
slow neutron and
a nucleus

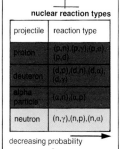

nuclear reaction types

projectile	reaction type
proton	(p,n),(p,γ),(p,α),(p,d)
deuteron	(d,p),(d,n),(d,α),(d,γ)
alpha particle	(α,n),(α,p)
neutron	(n,γ),(n,p),(n,α)

decreasing probability

carbon dating carbon–14 is produced in the atmosphere by the indirect action of cosmic rays, and atmospheric carbon dioxide contains a definite proportion of $^{14}_{6}C$. Non-living carbon-containing materials, e.g. wood, plant fibres, do not have this ^{14}C replaced by respiration, as in living organisms. The ^{14}C undergoes radioactive decay, and has a half-life of 5570 years; the proportion of ^{14}C to ^{12}C decreases with time and the age of the material can be estimated up to 50 000 years from the ^{14}C content.

nuclear reaction the reaction between two nuclei, or a nucleus and a positively charged particle or a neutron. Nuclei interact when they are close enough for nuclear attractive forces to overcome the electrostatic repulsion of positive charges; this repulsion does not apply to neutrons. The repulsive force between nuclei of atomic numbers Z and $Z-$ is $ZZ-e/d$, where e is the charge on an electron and d is the distance apart of the nuclei. The potential energy curve changes sharply at d_0 so that a minimum energy of $ZZ-e/d_0$ must be supplied to make the nuclei interact; this is $ZZ-$MeV approximately, and d_0 is approximately 10^{-14} m.

nuclear reaction types a nuclear reaction is brought about by bombarding a nucleus with a projectile. Common projectiles include protons, deuterons (↓), alpha particles and neutrons. The first artificial nuclear reaction was:
$$^{14}_{7}N + \alpha \rightarrow ^{17}_{8}O + p.$$
In abbreviated notation, this is $^{14}N(\alpha, p)$ ^{17}O. The main types of nuclear reactions, categorized by projectile, are shown in the table. The main types of nuclear reaction categorized by the effect on the nucleus are: elastic and inelastic scattering and transmutation reactions (↓).

deuteron (n) the nucleus of the deuterium atom, i.e. 1 proton and 1 neutron. Symbol for a deuteron is d.

compound nucleus a projectile combined with a nucleus.

transmutation reaction a nuclear reaction type (↑) in which the products differ from the reactants. *See* **nuclear fission** (p.64).

nuclear fragmentation a transmutation reaction (↑) in which the target nucleus acquires a very high excitation energy, so high that the target nucleus breaks rapidly into several large fragments.

fragmentation (n) *see* **nuclear fragmentation** (↑).

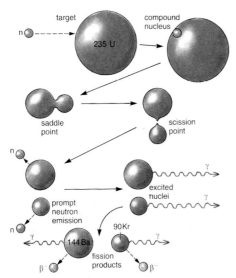

stages in nuclear fission of uranium-235

nuclear fission a transmutation reaction (p.63) in which a nucleus breaks up into two nuclei of comparable size together with a number of light particles (neutrons, protons, alpha particles). Fission may be spontaneous (↓) or induced (↓).

spontaneous fission fission of a nucleus in its ground state.

induced fission the nuclear fission (↑) of a compound nucleus formed from a bombarding particle and a heavy nucleus.

fission (*n*) *see* **nuclear fission** (↑).

fissile (*adj*) describes any isotope of an element that can undergo nuclear fission (↑), generally induced fission (↑), and is capable of practical use.

fissionable (*adj*) describes a nuclide which can undergo fission (↑) after capture of neutrons of all energies.

radiative capture a low energy neutron is absorbed into a nucleus and gamma radiation is emitted. Many commercially available radioactive isotopes use this reaction with thermal neutrons (p.65) available from nuclear reactors (p.66).

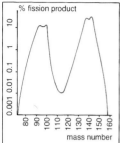

% fission product

mass number

fission products of uranium-235 by thermal neutrons

prompt neutron a neutron emitted as fission (↑) fragments fly apart. Neutrons emitted by stable isotopes, the products of fission, are 'delayed' neutrons. Prompt neutrons have a high energy.

thermal neutron a neutron with an energy on average equal to kT, where k is Boltzmann's constant and T is the temperature in kelvin. This is an energy of $8.6 \times 10^{-5}\,T$ electron-volts, obtained by slowing down high energy neutrons.

fission fragments a fissioning nucleus produces fission products dependent on the excitation energy and the compound nucleus (p.63) formed during fission. The fission products are radioactive isotopes of elements from the middle part of the periodic table, e.g. with mass numbers ranging from 72 to 160. Thermal neutron (↑) fission produces an asymmetric splitting of a nucleus into two unequal parts with a probability distribution of mass numbers of the nuclei. The number of prompt neutrons (↑) emitted depends on the fission products. For uranium–235, the weighted average is 2.5 neutrons per fission.

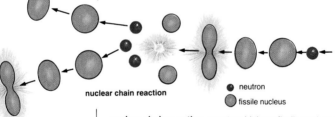

nuclear chain reaction

● neutron
● fissile nucleus

nuclear chain reaction a neutron hitting a fissile nucleus causes fission and the emission of a prompt neutron (↑). This neutron can then cause another fission, and so on, building up a fission chain reaction. If more than one neutron is emitted, the chain is extended rapidly and the total energy available from fission becomes enormous. Uranium–235 emits, on average, 2.5 neutrons per fission, and this isotope exists in natural ore deposits. The ore is stable against fission as most neutrons are lost in non-fission reactions with other nuclides (p.62), and some escape from the ore altogether. To maintain a chain reaction, an average of 1 neutron per fission must be available for further nuclear reaction.

fission chain reaction *see* **nuclear chain reaction** (↑).

critical mass in a specimen of pure uranium–235, many neutrons escape from the surface before they cause a fission. The fraction escaping can be reduced by increasing the volume, and hence the mass of the nuclide. There is a critical mass, with a definite value for each fissile nuclide, below which the nuclear chain reaction is not maintained. Above the critical mass, the chain reaction proceeds; the larger the mass the more rapid the chain reaction, until it becomes explosive.

moderator (*n*) a substance which slows down fast neutrons even to the speeds of thermal neutrons (p.65). Substances with low mass numbers, such as kerosene, water and graphite, make the best moderators. Energy is lost by inelastic scattering.

nuclear reactor a device for the controlled release of nuclear fission energy. It consists of an active core containing fissile material, with the fission controlled by a moderator (p.65) and control rods of a material, such as boron-steel or cadmium, which is a good absorber of neutrons. Nuclear fission develops heat, which is removed by a suitable coolant; substances used for this purpose include liquid sodium, carbon dioxide, helium. The coolant is used to generate steam which drives a conventional turbine. Control rods in the core are moved up and down to increase or decrease the multiplication factor (↓). The reactor is surrounded by a biological shield (p.58). The fast neutrons from atomic fission are slowed down to thermal neutrons (p.65) by the moderator; and the control rods keep the multiplication factor just slightly greater than one.

multiplication factor the number of neutrons which result from a single atomic fission. If the multiplication factor is less than one no chain reaction can take place; if it is greater than one a chain reaction may take place.

supercritical (*adj*) describes a nuclear fuel (↓), or an amount of fissionable material with a multiplication factor (↑) greater than one, i.e. a mass greater than the critical mass, or a volume greater than a critical volume.

nuclear fuel a fissionable nuclide used in a nuclear reactor (↑). Common fuels are uranium–238 enriched with uranium–235, and plutonium.

fertile material a material which can be converted into a fissile material, e.g. uranium–238 which can be converted into plutonium–239, or thorium–232 which can be converted into uranium–233.

atomic pile another name for a **nuclear reactor** (↑).

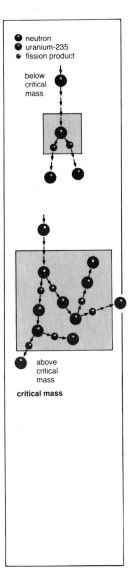

● neutron
● uranium-235
● fission product

below critical mass

above critical mass

critical mass

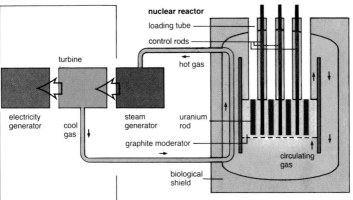

nuclear reactor
loading tube
control rods
turbine
hot gas
electricity generator
cool gas
steam generator
uranium rod
graphite moderator
circulating gas
biological shield

atomic energy the energy released in nuclear fission (p.64). The energy released from the fission of uranium–235 by thermal neutrons is shown in the table. The fission of a single uranium–235 atom releases approximately 200 MeV. By Einstein's equation $E = mc^2$, where c, the speed of light, is $2.998 \times 10^8 \, \text{ms}^{-1}$; if m is in kilograms, then E is given in joules. When mass is measured in amu, the energy in MeV is given by: $E(\text{MeV}) = m(\text{amu}) \times 931$, and an energy of 1 MeV per atom is equal to $9.65 \times 10^7 \, \text{kJ mol}^{-1}$. An energy release of 210 MeV per atom of uranium–235 is: $200 \times 9.65 \times 10^7 = 1.93 \times 10^{10} \, \text{KJ}$ for 235 g of uranium–235.

nuclear energy the correct term for **atomic energy** (↑).

fast reactor a nuclear reactor (↑) with no moderator. The emitted neutrons have an energy of 0.5 MeV or more; they cause nuclear fission.

thermal reactor a nuclear reactor (↑) with a sufficiently large moderator to reduce all emitted neutrons to thermal neutrons (p.65). The core of the reactor is surrounded by a reflector to scatter neutrons back and prevent them escaping from the core.

fast fission breeder reactor a fast reactor (↑) which uses plutonium–239 as a nuclear fuel. The central core is surrounded by uranium-238, which captures neutrons escaping from the core, and is converted to plutonium:
$$^{238}_{92}\text{U} + ^{1}_{0}\text{n} \rightarrow ^{239}_{92}\text{U} \rightarrow ^{239}_{94}\text{Pu} + 2\beta^-.$$
The reactor is called a breeder because it 'breeds' plutonium–239 from the uranium.

fast breeder another name for a **fission breeder reactor**.

advanced gas-cooled reactor a nuclear reactor (p.66) using uranium oxide enriched with uranium–235. The coolant is carbon dioxide and the reactor operates at higher temperatures than a thermal reactor (p.67). It has a graphite moderator.

high-temperature gas-cooled reactor a nuclear reactor (p.66) using uranium enriched with uranium–235 as a fuel. It has a graphite moderator, and operates at high temperatures. The nuclear fuel rods are encased in ceramic materials to avoid corrosion of metals. The coolant is helium.

nuclear fusion a nuclear reaction (p.63) in which two nuclei combine to form one heavier nucleus.

fusion (n) an alternative name for **nuclear fusion** (↑).

thermonuclear reaction nuclear fusion (↑) reactions can be brought about by bombarding targets of light elements with accelerated positive particles, but more energy is expended in producing the reaction than is gained from the reaction. If the temperature is raised to the order of 10^7 K, then the reaction is self-sustaining after an initial start, and a considerable quantity of energy is released. This is a thermonuclear reaction.

thermonuclear fusion the isotopes of hydrogen, deuterium and tritium offer a possibility of nuclear fission for obtaining energy on earth. Heavy water, D_2O, is bombarded by accelerated deuterons, the reaction is:

$$^2D + {}^2D \rightarrow {}^3_2He + n + 3.2\,MeV$$
$$^2D + {}^2D \rightarrow {}^3T + {}^1H + 4.0\,MeV$$

The high temperature required to initiate the reaction is obtained from nuclear fission. This is the basis of the thermonuclear, or hydrogen, bomb.

nuclear stability there is a narrow range of combinations of protons and neutrons which form stable nuclides. The graph shows this range. The reason for the narrow range is that the Pauli principle (p.87) demands a pair of protons and a pair of neutrons in a sub-shell, with each pair having opposing spin. Opposing spins release 'pairing energy' and form a stronger bond between the nucleons. This extra stability is clearly seen in the table for the distribution of stable nuclides for odd and even numbers of nucleons. Stable nuclei have more neutrons than protons; extra neutrons are required to counterbalance the repulsive force between protons in the nucleus.

nuclear spin nuclear angular momentum, symbol I.

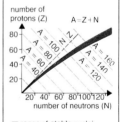

number of
protons (Z)   A = Z + N

80
60
40
20

20 40 60 80 100 120
number of neutrons (N)

■ range of stable nuclei

classification			number of stable nuclides
Z	N	A	
even	even	even	164
even	odd	odd	55
odd	even	odd	50
odd	odd	even	4

nuclear stability

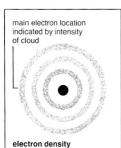

main electron location indicated by intensity of cloud

electron density

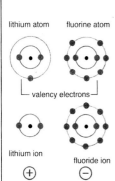

lithium atom   fluorine atom

valency electrons

lithium ion

fluoride ion

$\oplus$ $\ominus$

oppositely charged ions attract each other

ionic bond

chemical bond a force between two atoms which holds those atoms together as a result of the sharing or transfer of electrons (p.43) between the two atoms (p.29). Chemical reactions involve the breaking and making of chemical bonds as a result of the absorption or release of energy.

electron density refers to the concentration of the electron cloud at any particular distance from the nucleus (p.45) of an atom (p.29). It is the probability of finding the electron at that point.

electron density map a diagram, like a contour map, showing the most likely orbitals occupied by the electrons in an atom (p.29).

ionic bond a chemical bond (↑) formed between atoms as a result of one atom totally transferring one or more of its outer shell bonding electrons to the other atom. Such bonds are normally formed between a metal and a non-metal. In the process both atoms become ionized (p.125), the metal becoming a cation (p.126) and the non-metal an anion (p.126). As a result the actual chemical compound (p.16) consists of a uniform mixture of an equal number of positive and negative ions attracted to each other by their opposite charges forming the ionic bonds. Lithium bromide and potassium iodide are examples of compounds with ionic bonds.

electrovalent bond *see* **ionic bond** (↑).

electrovalency (*n*) refers to the number of ionic bonds an individual atom can form. It corresponds to the number of outer valency shell electrons that the atom can readily lose or gain in order to become an ion, e.g. a sodium atom can only lose one electron from its valency shell, so it has an electrovalency of $+1$. A bromine atom has 7 electrons in its valency shell and is able to gain one additional electron to form a negatively charged ion, so it has an electrovalency of -1. In both cases the atoms form ions with an outer electron shell containing eight electrons. This octet (p.70) of electrons is a very stable structure for the ion.

Fajans' rules Fajans specified two rules in order to judge whether the bond between two atoms would be an ionic bond (↑) or a covalent bond: (1) a covalent bond is more probable and an ionic bond less probable if the ions possess multiple charges; (2) a covalent bond is more likely if the atom produces a small cation (p.126) or a large anion (p.126).

stable octet refers to the tendency for all elements (except hydrogen and helium) to seek to form chemical bonds by sharing enough electrons in order to acquire an outer valency shell containing eight electrons. This stable octet is characteristic of the very stable electron configuration of the poorly reactive noble gases neon, argon, krypton and xenon. When a stable octet structure has been acquired the element cannot normally form any additional chemical bonds (p.69).

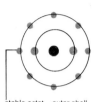

stable octet – outer shell with eight electrons
stable octet
e.g. neon atom

ionization energy the minimum energy required to remove the outermost (bonding) electron from an atom to form an ion (p.124). The symbol I is used for this term.

electron affinity (1) the energy released, measured in electron-volts, arising from the attachment of an electron to an atom to create a negative ion. The symbol E is used for this term. (2) a loose term for the general tendency an atom may have to attract electrons, thus the fluorine atom has a greater electron affinity than has the chlorine atom.

magnetic moment refers to the couple (force) needed in order to hold a magnet at right angles to a magnetic field of 1 ampere per metre $(1\,A\,m^{-1})$.

covalent bond one form of chemical bond (p.69) produced between atoms as a result of sharing electrons. Covalent bonds are obtained when the two atoms both retain partial control of the two electrons forming the bond. Such bonds may be formed as a result of one atom providing both electrons, *see* **coordinate bond** (p.77), or by the two atoms each providing a single atom to the bond. In all cases (except hydrogen) the two atoms will finish up with an outer electron configuration of an octet of electrons. Organic compounds (p.16) mainly have covalent bonds, with each carbon atom being capable of forming four such bonds either with other carbon atoms, or with elements such as hydrogen, oxygen, nitrogen and halogens. Covalent compounds have much lower melting points than do electrovalent (ionic) compounds.

hydrogen atoms forming a **covalent bond** by sharing their individual electrons

covalency (n) refers to the number of covalent bonds (↑) an atom is capable of forming. This corresponds to the number of electron orbitals in the atom which contain single electrons, e.g. nitrogen has three such unpaired electrons and is capable of forming three covalent bonds by sharing these with unpaired electrons from three other atoms.

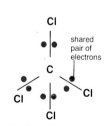

shared pair of electrons

carbon atom with a **covalency** of four

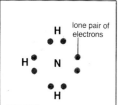

lone pair
e.g. ammonia molecule

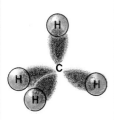

bond formation between
carbon and hydrogen
by **orbital overlap**

sigma bond

molecular orbital formed
from sharing two electrons
between carbon atoms

shared pair refers to the two electrons, one from each of two atoms, forming a covalent bond (↑) between the two atoms.

lone pair describes any two electrons filling a valence shell of an atom such that they cannot be shared separately to help form two individual covalent bonds (↑). They can, however, be donated together to form a single coordinate bond (p.77) with another atom requiring two electrons to complete an octet of electrons, e.g. two of the valence shell electrons of the nitrogen atom are a lone pair, as a result nitrogen can form one coordinate bond in addition to three normal covalent bonds.

orbital overlap a concept from molecular orbital theory in which bond formation is explained on the basis of the overlap of two electron orbitals, one from each atom. Each orbital is thus able to share its electron with the orbital of the other atom. By this means each orbital obtains a share in the electron of the other orbital; the bond is formed by the mutual sharing of the two electrons.

molecular orbital refers to the resulting orbital formed by the orbital overlap (↑) occurring when two atoms form a bond by sharing two electrons. The resulting molecular orbital may take several shapes depending upon the nature of the original electron orbitals. Thus a covalent bond (↑) formed from the overlap of two s-orbitals or an s- and a p-orbital will produce a strong linear sigma (σ) bond. But a covalent bond formed by parallel overlap of two p-orbitals will result in a weaker covalent pi (π) bond.

bonding molecular orbital a low energy state for a molecular orbital (↑) in which the bonding orbital is uniformly distributed between the two atomic centres.

anti-bonding molecular orbital a high energy activated state of a molecular orbital (↑) in which the bonding orbital is divided into two parts around the two atomic centres, the two orbitals effectively repelling each other because of the identical characteristics of the electrons occupying the orbitals.

molecular orbital theory the concept that molecular structures and chemical bond formation can be explained on the basis of the formation, overlap and combination of electron orbitals between atoms.

valence bond theory the concept that molecular structures and chemical bond formation can be explained on the basis of sharing pairs of electrons between atoms to form definite links between the atoms.

single bond a bond between two atoms formed by the sharing of just two electrons. This might be in the form of either a covalent bond (p.70) or a coordinate bond (p.77). The majority of chemical bonds are single bonds.

double bonds refers to the existence of two bonds between two atoms arising from the sharing of two sets of two electrons. Usually the two bonds will be of different strengths, one will be a strong sigma bond (↓) and the other a weaker pi bond (↓). In chemical reactions the pi bond will be the first to be broken. Compounds with double bonds are most common in organic chemistry.

triple bonds refers to the existence of three bonds between two atoms arising from the sharing of three sets of two electrons. The three bonds will usually consist of one sigma bond (↓) and two weaker pi bonds (↓). In chemical reactions the pi bonds will be the first to be broken. Triple bonds are fairly uncommon, the majority of them occurring in organic compounds.

catenation (*n*) refers to the tendency of the atoms of an element to join together to form chains. Carbon is well known for this, with its wide range of hydrocarbons (p.34) and other homologous series (p.35), but other elements such as boron and germanium will form short chains. Catenation also occurs with two or more elements forming chains, e.g. silicon and oxygen.

sigma bond often written as σ-bond, it refers to a covalent bond (p.70) formed as a result of the orbital overlap (p.71) of two s-orbitals or an s- and a p-orbital. *See also* **molecular orbital** (p.71).

pi bond often written as π-bond, it refers to a covalent bond (p.70) formed as a result of the orbital overlap (p.70) of two parallel p-orbitals. These bonds are weaker than sigma bonds and occur in unsaturated compounds such as ethene and ethyne.

delocalized pi bond refers to a pi bond (↑) spread over the bond lengths of several linked carbon atoms due to them all contributing p-orbitals to the pi bond system. As a result the pi bond cannot be considered to be localized between any two carbon atoms. Butan–1,3–diene has a delocalized pi bond system.

mobile electron describes an electron which may move readily from one atom to another in a chemical structure depending upon the external chemical environment. Reaction mechanisms involve the movement of mobile electrons around molecules (p.29) and ions (p.124).

cyclohexene – a compound with a carbon-carbon **double bond**

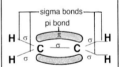

benzonitrile – a compound with a carbon-nitrogen **triple bond**

the ethene (ethylene) molecule showing **sigma bonds** and **pi bond**

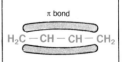

delocalized pi bond in the 1,3-butadiene molecule

hybridization (*n*) a process by which an atom forms two or more equivalent orbitals from atomic orbitals of different types. The resulting hybrid orbitals are at a lower energy level than the highest energy level of any of the original orbitals. Many types of hybridization are known, and it explains how an atom (p.29) can form a number of equivalent chemical bonds. **hybrid** (*n*), **hybridized** (*adj*).

contributing orbitals			hybrid form
one s,	*three p,*	*two d*	*six sp^3d^2*
one s,	three p,	one d	five sp^3d
one s,	three p,		four sp^3
one s,	two p,		three sp^2
one s,	one p,		two sp

degeneracy (*n*) describes atomic and molecular energy levels within a single system which are identical, i.e. the energy levels are accidentally coincidental. **degenerate** (*adj*).

bond energy the energy released when a covalent bond (p.70) is formed between two atoms. It may be considered as a measure of the strength of the bond, e.g. the bond energy for the carbon-oxygen bond C–O is 354kJ mol^{-1}.

bond dissociation energy refers to the average energy in kJ required to bring about fission of a chemical bond (p.69), at a temperature of 25°C, e.g. for a carbon-hydrogen bond C–H in methane the bond dissociation energy is 406kJ.

bond length the mean internuclear distance between two atoms forming a chemical bond (p.69). Even for the same atomic species the bond length will differ from one compound to another, for example the carbon-hydrogen bond length in benzene differs from that in cyclohexane.

covalent radius refers to half the distance between the two nuclei (p.45) form a molecule (p.29) of an element (p.16). This is less than the radius of the atom (p.29) due to the orbital (p.49) overlap arising from the bond formation.

bond angle the angle between any two chemical bonds (p.69) in a compound. It is measured as the angle between the lines joining the centres of the three nuclei forming the two bonds, e.g. the bond angle in water is 105° and for methane all the bond angles are 109°5'.

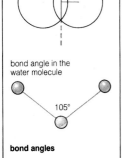

bond length

bond	length (nm)
C — C	0·153
C — H	0·110
O — H	0·096
N — H	0·101
H — H	0·074
F — F	0·142
H — Cl	0·127
H — F	0·092

covalent radius

r

bond angle in the water molecule

105°

bond angles

molecular structure a molecule is formed as a result of chemical bonds (p.69) between atoms. Each of these bonds is in a fixed direction with respect to the other bonds in the molecule. The molecular structure is the three-dimensional shape determined by the bond angles (p.73) and bond lengths (p.73) between the atoms. The most important molecular shapes are linear, pyramidal, tetrahedral and octahedral. Organic molecules often have very complex molecular structures.

linear molecule any molecule in which the atoms are joined in a straight line, e.g. dinitrogen monoxide N_2O.

non-linear molecule a simple molecule in which the bonds between three atoms are at an angle to each other and not in a straight line, e.g. water H_2O in which the bond angle (p.73) is 105°.

trigonal planar molecule a molecule, or ion (p.124), in which a central atom is bonded to three other atoms and all four atoms lie in the same plane, e.g. sulphur trioxide SO_3, the nitrate ion NO_3^-.

tetrahedral molecule any molecule in the shape of a tetrahedron, in which a central atom has four valence bonds (p.71) directed towards the four points of the tetrahedron. Tetravalent elements such as carbon and germanium will form tetrahedral molecules. The bond angles (p.73) for the four bonds in a tetrahedral molecule are ideally 109°5′, e.g. methane CH_4, carbon tetrachloride CCl_4.

octahedral molecule a molecular shape in the form of an octahedron in which a central atom has six chemical bonds (p.69) directed to the points of an octahedron, e.g. sulphur hexafluoride, SF_6, has an octahedral molecule.

pyramidal molecule any molecular shape in the form of a pyramid in which the central atom has either three or four chemical bonds (p.69) directed to other atoms forming the base of the pyramid, the molecules may be trigonal pyramidal (↓) or square pyramidal (↓).

square pyramidal molecule a pyramidal molecule (↑) in which a central atom (the peak of the pyramid) is bonded to four other atoms forming the corners of a square base pyramid, e.g. $\pm F_5$.

trigonal pyramidal molecule a pyramidal molecule (↑) in which a central atom (the peak of the pyramid) is bonded to three other atoms forming the corners of a triangular base to the pyramid. The ammonia molecule has the trigonal pyramid shape.

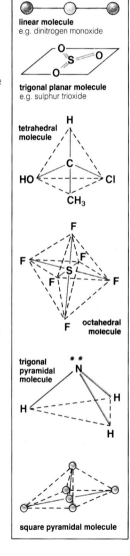

linear molecule
e.g. dinitrogen monoxide

trigonal planar molecule
e.g. sulphur trioxide

tetrahedral molecule

octahedral molecule

trigonal pyramidal molecule

square pyramidal molecule

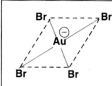

square planar molecule

$C_2H_5 - O - H$
$\delta^- \quad \delta^+$ (over O and H)

polar molecules
polar hydroxyl group
in ethanol

square planar molecule refers to molecules in which the central atom is bonded to four other atoms in the same plane and positioned at the corners of a square. The precious metals gold and platinum give ions (p.124) which have the square planar structure, e.g. $AuCl_4^-$ and $AuBr_4^-$.

inductive effect a permanent effect in carbon chain compounds in which electrons forming a bond between a carbon atom and a non-carbon atom are partially displaced towards the atom with the greatest electronegativity (↓). This effect gets progressively weaker along a carbon chain.

electronegativity (n) the tendency for an atom forming a bond to attract the electrons constituting the bond. In a compound such as hydrogen fluoride the electrons forming the bond between the two atoms are attracted towards the fluorine atom as its electronegativity is nearly double that of the hydrogen atom. Elements (p.16) can be presented in terms of an electro-negativity scale, fluorine having a value of 4.0 and hydrogen 2.1.

dipole moment the product of the numerical value of one of two equal and opposite charges and the distance between the two charges. Many molecules possess dipole moments if electrons (p.43) are unevenly shared between atoms, e.g. as in polar groups such as hydroxyl (p.36), carboxyl (p.36) and ketone (p.39) groups. Dipole moments are measured in debye units (↓).

polar molecules refers to covalent molecules possessing dipole moments (↑) due to electrons forming a bond being attracted closer to one atom rather than the other. As a result one atom is slightly positive whilst the other is slightly negative (this polarity of the bond is shown by + and − respectively on the two atoms), for example in alcohols (p.39) the hydroxyl group (p.36) is polar as shown in the diagram.

non-polar molecules (1) molecules which do not possess dipole moments (↑) as they are not polar, e.g. O_2, H_2, Cl_2, N_2; (2) molecules which are symmetrical and possess a uniform charge distribution, e.g. benzene, tetrachloromethane.

debye unit a unit for expressing the values for the dipole moments (↑) of molecules, one $D = 3.33563 \times 10^{-30}$ coulomb-metre.

polar liquid a liquid consisting of molecules (p.29) possessing dipole moments (p.75). Such liquids are capable of dissolving ionic compounds. e.g. water is a polar liquid with a dipole moment of 1.84 and is an excellent solvent for a wide range of substances.

polar solvent a polar liquid (↑) used to form solutions by dissolving other substances.

non-polar solvent any liquid of non-polar molecules (p.75) employed as a solvent for other substances. Such liquids are usually good solvents for compounds with low polarities.

polarizing power a measure of the power of one atom or ion to polarize (dislocate) the bond between itself and another atom or ion. For a bond between two atoms, it is the atom with the greater electronegativity (p.75) which polarizes the bond giving rise to any dipole moment. The degree of distortion caused to the covalent bond (p.70) depends upon the polarization of the polarized atom. In the case of two ions, it is the electronegativity of the ion which determines the magnitude of the polarizing power. Cations (p.126) with large charges and small radii possess high polarizing powers, and this provides an explanation for Fajans' rules (p.69).

molar polarization a relationship between the dielectric constant (p.11) of a substance and its density at a specified temperature given by the equation:

$$P_M = \frac{\varepsilon - 1}{\varepsilon + 2} \frac{M}{\rho}$$

where ε is the dielectric constant, M is the molar mass, ρ is the density. For non-polar molecules the value of P_M does not change with temperature, whilst for polar molecules it decreases as the temperature increases.

conjugative system any molecular system in which a structure of alternate single and double bonds (p.72) interacts in a manner which leads to delocalization of the double bond electrons. The delocalization creates a more stable, low energy system, e.g. buta–1,3–diene and benzene. **conjugate** (*v*), **conjugation** (*n*).

resonance (*n*) the existence of a molecular structure in a form which is best represented as being intermediate between two or more simple structures, e.g. benzene is considered to form a resonant hybrid (↓) intermediate between five canonical (↓) forms. Its true structure shows a bond character which is neither a simple single bond (p.72) nor a double bond (p.72).

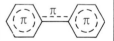

conjugative system
delocalization of the π electron clouds in the conjugation of diphenyl

resonance energy the difference between the heat of formation of a resonance hybrid (↓) and any of the canonical structures (↓) considered to contribute to that hybrid. This difference is a measure of the increased stability of the resonant hybrid compared to the contributing structures.

resonance hybrid another term for the true molecular structure of a compound which cannot be represented in a single, simple diagrammatic form but only as something intermediate between a number of simple structures. See **resonance** (↑).

canonical structure refers to any of the specified structures contributing to a resonance hybrid (↑), e.g. carbon dioxide has three canonical forms:

$$O = C = O \longleftrightarrow \bar{O} - C \equiv \overset{+}{O} \longleftrightarrow \overset{+}{O} \equiv C - \bar{O}$$

canonical structure
the five canonical forms of benzene

mesomeric effect refers to localized concentrations of electric charge occurring within organic molecules as a result of resonance (↑) within a molecule.

mesomerism (*n*) another term for **resonance** (↑).

coordinate bond a form of covalent bond (p.70) in which one atom contributes both electrons forming the bond between it and the other atom. The electrons used for the bonding are a lone pair (p.71) on the first atom. The bond is often represented as a short arrow between the two atoms from the donor (↓) atom to the acceptor (↓) atom, A→B. Phosphorus and sulphur both form compounds involving coordinate bonds.

coordinate link another name for a **coordinate bond** (↑).

dative bond another name for **coordinate bond** (↑).

semipolar bond another name for **coordinate bond** (↑).

donor (*n*) the atom contributing the lone pair (p.71) to form a coordinate bond (↑).

acceptor (*n*) the atom accepting the lone pair (p.71) when a coordinate bond (↑) is formed.

coordination compound (1) any chemical compound containing one or more coordinate bonds. (2) a complex compound in which a number of coordinate bonds (↑) are formed between a ligand (p.78) and a metal, creating ring structures by **chelation** (p.208).

phosphoric acid

phosphorus atom contributing two electrons to form a **coordinate bond** with oxygen

coordination compound
e.g. copper (II) acetylacetonate

coordination number specifies the number of coordinate bonds (p.77) that can be formed between a metal and ligands (↓), e.g. copper has a coordination number of 4 as it forms an ion by coordination with four ammonia molecules, and iron has a coordination number of 6 as it forms an ion with six cyanide ions.

ligand (*n*) a chemical group bound to a central metal atom by a coordinate bond (p.77), the term is used particularly in connection with short chain (p.33) compounds possessing polar groups at each end of the chain with lone pairs (p.71) of electrons on the atoms, e.g. ethylene diamine is a ligand that can form coordinate bonds due to the lone pairs on both nitrogen atoms.

ligand nomenclature ligands are classified according to the number of coordinate bonds that can form with metals and may be referred to as monodentate (↓), bidentate (↓), etc.

monodentate ligand refers to a ligand (↑) capable of forming only one coordinate bond (p.77) with a metal, e.g. water H_2O, and ammonia NH_3 are monodentate ligands as they will form a bond with a metal atom or ion at only one point by donation of a lone pair (p.71) of electrons to the metal.

bidentate ligand a ligand (↑) molecule possessing two atoms each with a lone pair (p.71) which can act as donors (p.77) to form coordinate bonds (p.77) with metals, e.g. ethylene-diamine is a bidentate ligand.

tridentate ligand a ligand (↑) molecule possessing three atoms each with a lone pair (p.71) capable of being donated to form coordinate bonds (p.77) between the ligand and metal atoms or ions, e.g.

(den): $NH(CH_2CH_2NH_2)_2$.

copper with a **coordination number** of four in the cuprammonium ion

$$[Cu(NH_3)_4]^{2+}$$

ethylenediamine – a molecule which acts as a **ligand** in forming coordinate bonds with metals

$$H_2N - CH_2 \ CH_2 - NH_2$$

monodentate ligand

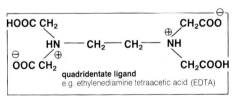

quadridentate ligand
e.g. ethylenediamine tetraacetic acid (EDTA)

quadridentate ligand a ligand (↑) molecule possessing four atoms each with a lone pair (p.71) which can act as donors (p.77) to form coordinate bonds (p.77) with metals, e.g. ethylenediaminetetraacetic acid (EDTA).

$[Cr(H_2O)_6]^{3+}\ 3Cl^-$

$[ClCr(H_2O)_5]^{2+}\ 2Cl^- . H_2O$

$[Cl_2Cr(H_2O)_4]^+\ Cl^- . 2H_2O$

the three **ionization isomers** of chromium (III) chloride hexahydrate

hexadentate ligand a ligand molecule possessing six atoms each with a lone pair (p.71) which can act as donors (p.77) to form coordinate bonds (p.77) with metals,
e.g. penten $(H_2NCH_2CH_2)NCH_2CH_2N(CH_2CH_2NH_2)_2$.

hydrate (*n*) a chemical compound in which molecules of water are associated with a central metal ion (p.124) as a result of coordinate bonds (p.77) formed by the donation of the lone pairs (p.71) of electrons from the oxygen atoms to the metal, e.g. copper (II) sulphate has five molecules of water combined with it to give the molecule represented as $CuSO_4.5H_2O$. **hydrated** (*adj*). **hydration** (*n*).

ionization isomers compounds with identical molecular formulae but which exist in different ionic forms. These are usually hydrated (↑) salts of inorganic compounds with different coloured ions, e.g. chromium (III) chloride hexahydrate exists in three ionic forms each with the same molecular formula.

metallic bond the form of bond occurring in metals and metal alloys in which a regular arrangement of positively charged metal ions (p.124) is held together by free electrons (↓).

free electrons electrons in a crystal (p.165) or metal lattice which are free to move under external influences. When a potential difference is applied across the material the free electrons will move to give an electric current.

energy bands refers to continuous bands of energy arising from a series of close energy levels occurring mainly in metals where the valency electrons form bonds with an increased number of quantized energy levels.

energy bands in different materials

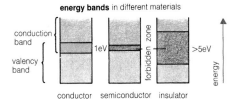

conduction band

valency band

forbidden zone

1eV

>5eV

energy

conductor   semiconductor   insulator

valency band a continuous band of quantized energy levels arising from the valency electrons of atoms, usually metals, in a crystalline solid.

conduction band a high energy band of quantized energy levels to which electrons (p.43) may be elevated as a result of thermal agitation of the crystal lattice (p.165). An electron on entering a conduction band leaves a vacancy, a hole, in the lower energy valency band (p.79). Once in the conduction band the now free electrons (p.79) are mobile and will move when a potential difference is applied.

conductors (*n*) substances consisting of materials which readily permit electricity and/or heat to flow along them. Most good conductors are metals or solutions of electrolytes. **conduction** (*n*), **conduct** (*v*).

insulators (*n*) substances consisting of materials which do not easily permit electricity and/or heat to pass through them. Ceramic materials and organic polymers are good insulators as they do not possess free electrons (p.79) or metallic bonds (p.79).

semi-conductors (*n*) substances, usually made synthetically, which possess electronic properties which are intermediate between those of conducting metals and insulating non-metals. Properties of semi-conductors change with temperature, their conductivity increasing due to heat or irradiation with light. They are prepared from specially treated crystals of elements such as germanium, silicon, selenium and tellurium.

hydrogen bond a weak bond formed between hydrogen atoms in compounds and strong electronegative (p.75) atoms in other molecules. Oxygen, fluorine, and nitrogen are most commonly associated with hydrogen bonds, the bond being represented as a dotted line between two elements. The hydrogen bonds between water molecules are responsible for the exceptional physical properties water possesses as they lead to high boiling points and high melting points.

H   H
| hydrogen bond |
H————O :‐ ‐ ‐ ‐ ‐ ‐ ‐ ‐ ‐H————O :

intermolecular **hydrogen bond** between molecules of water

association (*n*) refers to molecules which are attracted to each other usually due to weak forces such as those creating hydrogen bonds (↑) or due to dipole-dipole attraction (↓).

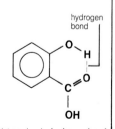

intramolecular **hydrogen bond** in salicylic acid

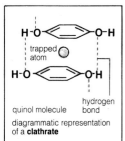

quinol molecule   hydrogen bond

diagrammatic representation
of a **clathrate**

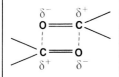

dipole – dipole attraction
due to polarization of
electron charge in carbonyl
groups

clathrates (*n*) a group of chemical substances in which
small molecules or atoms of one species are trapped
inside a cage-like structure formed from the crystals
(p.165) of another substance. e.g. urea forms a series
of such compounds with the alkanoic (p.34) acids. and
the noble gases can similarly be trapped inside quinol
crystals. There are no bonds between the crystals and
the trapped molecules.

host structure the crystal structures forming the cage
which traps the guest molecules (↓) in clathrates (↑).

guest molecule refers to the molecules trapped
inside the crystal (p.165) host structure (↑) to form
clathrates (↑).

dipole-dipole attraction refers to the attraction between
adjacent molecules arising from the uneven distribution
of electrons (p.43) in functional groups (p.36) which
creates a dipole in that part of the molecule in
which one end of the group possesses a small
positive charge whilst the other end has a small
negative charge.

dipole induced dipole force refers to a distortion or
polarization of the electrons forming bonds in
molecules leading to the creation of an induced dipole
as a result of the effects of other nearby molecules
already possessing dipoles.

dispersion force a very weak force occurring between
adjacent molecules as a result of the mutual attraction
due to opposite charges of electrons (p.43) and nuclei
of the atoms in the adjacent molecules.

van der Waals' bond a weak bond occurring between
molecules due to the movement of electrons and nuclei
(p.45) producing a form of weak oscillating dipole
giving a degree of attraction between molecules. These
forces are much weaker than are those between ionic
substances.

intermolecular force a collective term for the attractive
and repulsive forces occurring between atoms and
molecules due to the presence of separated positively
charged nuclei (p.45) and negatively charged
electrons (p.43) giving rise to polarized groups,
dipoles, van der Waals' forces (↑) and hydrogen
bonds (↑).

van der Waals' radius refers to the effective distance
around an atom or group (p.51) at which the forces may
be considered to operate which create van der Waals'
bonds (↑).

electromagnetic waves a form of energy that travels in straight lines and has the properties of a wave motion, i.e. the waves can be reflected and refracted and can suffer interference (↓). Electromagnetic waves have an electric vector and a magnetic vector at right angles to each other, and both vectors are at right angles to the direction of travel. The waves are transverse in nature, i.e. variations of the electric and magnetic properties are at right angles to the direction of travel. They do not need a material medium as they can travel through space. The speed in space of all electromagnetic waves is approximately $3.0 \times 10^8 \, ms^{-1}$ and this is the highest speed that can be reached in the universe. The speed of the waves is reduced when they pass through electromagnetically dense materials; this reduces the wavelength (↓), but the frequency (↓) remains constant. Penetration of a medium depends on the frequency of the waves. The symbol for the speed of electro-magnetic waves is c.

wavelength (n) the distance between a point in a wave and the next point at the corresponding place moving in the same direction, i.e. the distance between one crest and the next crest. The symbol for wavelength is λ.

frequency (n) the number of crests that pass a point in one second. Frequency, wavelength and speed of the wave motion are connected in one equation $c = \nu\lambda$, where ν is the symbol for frequency.

spectrum (n) a display of different frequencies (alternatively of different wavelengths) over a certain range. There are several different types of spectra.

electromagnetic spectrum (e.m.) the range of frequencies (or wavelengths) of electromagnetic waves (↑). It ranges from the highest frequencies in gamma radiation to the lowest frequencies in radio waves.

visible spectrum the part of the electromagnetic spectrum that is sensed by the eye. The different frequencies (or wavelengths) are seen as colours. If white light is passed through a glass prism, it splits into the colours of the spectrum. The frequencies and wave-lengths associated with each of the colours are shown.

spectrometer (n) an instrument for measuring the wavelengths, and hence the frequencies of light in the visible spectrum. It consists of a **collimator**, glass prism, and **telescope**. The collimator has a slit, through which light enters, and is focused into a parallel beam of light. The prism splits white light into a spectrum.

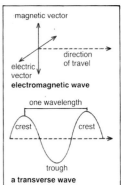

electromagnetic wave

a transverse wave

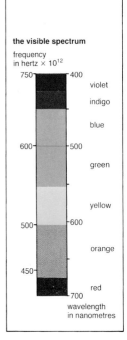

the visible spectrum

plan of a spectrometer

**emission spectrum
for sodium**

The spectrum is seen through a telescope. A scale on the circular table measures angles between the collimator and the telescope, and from these angles the wavelength of any e.m. wave can be found. Prisms of other materials can be used for ultraviolet and infrared spectra, and a heat detector or a photocell is used instead of viewing the spectrum by eye.

intensity (n) the quantity of energy carried by a wave motion in one second through a unit area. Intensities can be compared by means of a Geiger counter (p.57) for X-rays (p.92) and gamma rays (p.53), a photocell for u.v. rays and light, and a heat detector for i.r. radiation.

interference (n) the interaction between two sets of waves when both pass through the same space; the waves must be of the same frequency and from the same source. Where the crest of one wave meets the trough of another wave, both waves are destroyed. If two crests or two troughs meet, the resulting wave is bigger. Interference produces alternate bands of high intensity (↑) and low intensity.

ultraviolet radiation electromagnetic waves with wavelengths shorter than those of the visible spectrum (↑). The u.v. spectrum is continuous with the violet end of the visible spectrum. Measurements of wavelengths are made using a quartz prism in a spectrometer, and detecting the waves by a photographic film. Ultraviolet radiation has a chemical effect on many substances. *See* **electromagnetic spectrum**.

infrared radiation electromagnetic waves with wavelengths longer than those of the visible spectrum (↑). The i.r. spectrum is continuous with the red end of the visible spectrum. Measurements of wavelengths are made using a rock salt prism for longer wavelengths, and a glass prism for shorter wavelengths, in a spectrometer. The radiation is detected by a thermistor. *See* **electromagnetic spectrum.**

emission spectrum a spectrum formed by a material which is heated, e.g. a white-hot wire, or by excitation from an electric arc or electric discharge, e.g. as seen in gases in a discharge tube. A spectrum can be a continuous line or a band spectrum. It is formed in the visible region or in the infrared or ultraviolet region of e.m. waves. It is produced by extranuclear electrons first being excited by energy, and being raised to a higher energy level, and then falling back to a lower energy level and emitting energy in the form of e.m. waves.

line spectrum a spectrum formed by the atoms of elements in an incandescent gas; it is a form of emission spectrum (p.83). Line spectra are also formed when salts of metals are heated in a flame. The spectrum consists of a series of lines.

band spectrum a spectrum formed by the molecules of compounds. It consists of broad bands of frequencies with a sharp edge at one end, gradually dying away at the other end. Close examination shows the bands are made up from very many fine lines.

continuous spectrum a spectrum in which frequencies change continuously from one end of the spectrum to the other, and no frequencies are missing.

absorption spectrum a spectrum produced by viewing white light through a specimen and then observing the spectrum by a spectrometer (p.82). Some of the wavelengths are absorbed by the specimen and black lines, i.e. lack of light, appear on a continuous spectrum. Absorption spectra are found in the visible and u.v. regions of e.m. waves. The specimen is either a vapour in a vessel or a flame coloured by a metal salt.

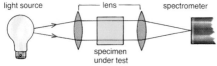

obtaining an absorption spectrum

ionization energy the least amount of energy required to remove one electron from an atom. It can be measured from line spectra. Lines in a line spectrum get closer together as the frequency increases, until they reach a limit, a measure of the ionization energy, *see* **Lyman series** (p.88).

wave number the number of crests of a wave in 1 metre. The symbol is σ; $\sigma = 1/\lambda$; the unit is metre^{-1}. For e.m. radiation $c = \nu\lambda$, so $\sigma = 1/\lambda = \nu/c$, i.e. $\nu = c\sigma$, i.e. frequency = (speed of e.m. waves) × (wave number).

Rydberg constant a constant which relates the wave number of spectral lines of atomic spectra for series which are similar to hydrogen. The Rydberg constant has the symbol R. The Rydberg formula is: $\sigma = R\,(1/n^2 - 1/m^2)$ where σ is the wave number of a spectral line, n and m are two integers such that $m > n$,

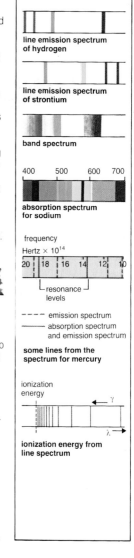

line emission spectrum of hydrogen

line emission spectrum of strontium

band spectrum

400   500   600   700

absorption spectrum for sodium

frequency
Hertz × 10^{14}

20   18   16   14   12   10

resonance levels

– – – – emission spectrum
———— absorption spectrum and emission spectrum

some lines from the spectrum for mercury

ionization energy

γ

λ

ionization energy from line spectrum

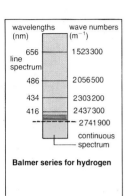

wavelengths (nm)	wave numbers (m^{-1})
656	1 523 300
486	2 056 500
434	2 303 200
416	2 437 300
	2 741 900

line spectrum

continuous spectrum

Balmer series for hydrogen

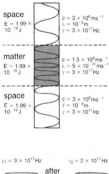

electromagnetic wave passing through matter. A quantum of energy for the radiation

space
$E = 1.99 \times 10^{-16}$ J
$c = 3 \times 10^8$ ms^{-1}
$\lambda = 10^{-9}$ m
$\gamma = 3 \times 10^{17}$ Hz

matter
$E = 1.99 \times 10^{-16}$ J
$c = 1.5 \times 10^8$ ms^{-1}
$\lambda = 5 \times 10^{-10}$ ms^{-1}
$\gamma = 3 \times 10^{17}$ Hz

space
$E = 1.99 \times 10^{-16}$ J
$c = 3 \times 10^8$ ms^{-1}
$\gamma = 10^{-9}$ m
$\gamma = 3 \times 10^{17}$ Hz

$\gamma_1 = 3 \times 10^{17}$ Hz $\gamma_2 = 2 \times 10^{17}$ Hz

photon → after collision → photon

change in momentum after collision

$= \frac{h\gamma_1}{c} - \frac{h\gamma_2}{c} = \frac{h}{c}(\gamma_1 - \gamma_2)$

$= \frac{6.625 \times 10^{-34}}{3 \times 10^8} \times (3 \times 10^{17} - 2 \times 10^{17})$

$= 2.208 \times 10^{-42} \times 1 \times 10^{17}$

$= 2.208 \times 10^{-25}$

units of momentum

e.g. in the Balmer series, $n = 2$, and $m = 3, 4, 5 \ldots \infty$. Each value of m produces a spectral line. When $m = \infty$, $\sigma = R/4$ for the Balmer series. This is the value for the ionization energy for this series.

quantum theory Planck put forward the hypothesis that the energy of electromagnetic radiation is emitted or absorbed in multiples of small amounts, that is, e.m. energy is discontinuous (as also is matter). The smallest amount of energy was called a **quantum**. Evidence for this hypothesis can be seen in the lines of atomic spectra, where the wave numbers of spectral lines are in agreement with Rydberg's formula (↑). Compare **electric charge** where the charge on an electron is 1.6×10^{-19} coulomb, the smallest charge that exists.

quantum (*n. pl. quanta*) *see* **quantum theory** (↑).

quantized (*adj*) describes a physical quantity that exists only in discrete units, i.e. it is not continuous, e.g. matter, energy, electric charge are quantized.

de Broglie's equation a particle in motion has associated with it the spread of energy by a wave motion. A particle of mass, m, moving with a velocity, v, has an associated wavelength (λ) given by de Broglie's equation: $\lambda = h/mv$ where h is Planck's constant (↓).

photon (*n*) a quantum (↑) of electromagnetic radiation. A photon can be thought of as a particle (p.43) associated with a wave motion. Its properties are: (a) a constant velocity of 3×10^8 ms^{-1}; (b) no electric charge; (c) a zero rest mass; (d) a change in momentum by altering the frequency associated with it. The change in momentum $= h v/c$ where v is the *change* in frequency and c is the speed of the radiation. A more restricted meaning makes a photon equal to a quantum of light energy.

Planck's constant the constant shows the relation between the frequency of a radiation and the magnitude of a quantum (↑) of energy. The symbol for Planck's constant is h. The accepted value for h is 6.625×10^{-34} J s. For e.m. radiation of frequency 5.069×10^{14} Hz, a quantum of energy is $5.069 \times 10^{14} \times 6.625 \times 10^{-34} = 3.37 \times 10^{-19}$ J. In general, $E = h v$ and hence $E = hc\sigma$, where σ is a wave number.

quantum mechanics a system of mechanics for atoms and molecules which uses the quantum theory (↑) when describing forms of energy. Quantum mechanics provides an explanation of the line spectra of atoms and the spectra of molecules.

wave mechanics a form of quantum mechanics in which extra-nuclear electrons are given the properties of three-dimensional wave forms, and not those of a particle. The wave nature and particle nature of an electron are connected by de Broglie's equation (p.85). Wave mechanics describes the shape and orientation of orbitals.

quantum number a whole number or one half. The whole number describes a number of quanta. Each extra-nuclear (p.45) electron in an atom is described by four quantum numbers; principal (↓), azimuthal (↓), magnetic (↓) and spin (↓) quantum numbers.

principal quantum number the quantum number that defines the electron shell (p.45) of an atom. Its symbol is n. The K-shell has $n = 1$; the L-shell has $n = 2$; the M-shell has $n = 3$, etc.

azimuthal quantum number the quantum number that defines the shape of the orbital of an electron; it also defines the sub-shells. Its symbol is l. An s-orbital (p.49) has $l = 0$; a p-orbital has $l = 1$; a d-orbital has $l = 2$.

magnetic quantum number the quantum number that defines the orientation of an orbital in a magnetic field. An electron in a p-orbital or higher orbital has a magnetic moment due to the motion of the electron in the orbital. The number of orientations is limited by quantum restrictions, and these are defined by the magnetic quantum number. Its symbol is m. The quantum restrictions are given by the whole numbers in the range $-l \leqslant m \leqslant +l$, i.e. if $l = 2$, then m can take the values of $-2, -1, 0, +1, +2$.

Zeeman effect when an external perpendicular magnetic field is applied to a discharge tube, the lines in a spectrum are split into 3, 5, 7 or more lines. This is known as the Zeeman effect. The orientation of the electron orbitals produces different energy levels in the magnetic field, and hence different spectral lines. Without the magnetic field, all the orbitals have the same energy level.

spin quantum number an electron in its orbital spins, and the spin can have two possible directions, clockwise and anticlockwise. Slightly different energies of the electrons arise from the possible spins. A spin quantum number, symbol s, is given to these spins; it can have the values $+\frac{1}{2}$ or $-\frac{1}{2}$. An orbital can be occupied by two electrons with opposite spins, but not by two electrons with the same spin. Electrons with opposite spins in an orbital are called an electron pair.

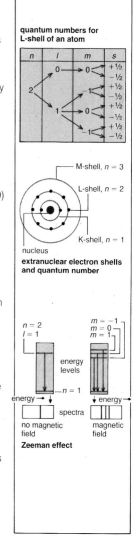

quantum numbers for
L-shell of an atom

extranuclear electron shells
and quantum number

Zeeman effect

relation of quantum numbers the values of l are 0 to $(n - 1)$, i.e. if $n = 3$, then $l = 0, 1, 2$. The values of m are from $-l$ to $+l$; s is always $+\frac{1}{2}$ or $-\frac{1}{2}$. For an L-shell, $n = 2$, so $l = 0$, $m = 0$, and $l = 1$, $m = -1, 0, +1$, i.e. 4 values in all; for each of these there are two possible values of s, hence 8 possible values, in agreement with the 8 electrons in the shell.

Pauli's exclusion principle each electron in a neutral atom is characterized by a unique set of four quantum numbers (↑), i.e. no two electrons can have the same set of quantum numbers $(n, l, m, s,)$.

Hund's rule states that electrons, as far as possible in accordance with Pauli's principle (↑), will go into orbitals so that they have parallel spins (↓). This means that the most stable electron configuration (p.88) in a sub-shell is one with the greatest number of unpaired electrons possible in accordance with Pauli's principle.

parallel spin two electrons have parallel spin, when both spin in the same direction, i.e. both of them have the same spin quantum numbers (↑). Electron spins are either *parallel* or *opposite*. There is a small lowering of energy for parallel spins compared with opposite spins.

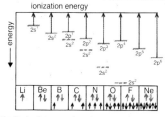

the ionization energies of one electron for atoms in the second period of the periodic table

↑ s electrons
↟ p electrons
↑↑ parallel spins
↑↓ opposite spins

principal quantum number	orbital
1	s
2	s → p
3	s → p → d
4	s → p → d → f
5	s → p → d → f
6	s → p → d → f
7	s → p → d

order of filling atomic orbitals

Aufbau principle the orbital occupied by an additional electron is decided by the nature of the available orbits and is such as to make the energy of the whole system a minimum. Hence an additional electron enters the next lowest available energy level orbital. *See diagram* for the order of filling atomic orbitals; this gives the order for the n and l quantum numbers. *See diagram* for the ionization energy of elements in group 2 of the periodic system. Atoms with the greatest ionization energies are the stablest. Notice that nitrogen with 3 parallel spin electrons is more stable than oxygen with one electron pair and two parallel spin electrons.

n	1	2				
l	0	0	1			element
m	0	0	+1	0	−1	
	↑ ↓					He
	↑ ↓	↑				Li
	↑ ↓	↑ ↓	↑	↑		C
	↑ ↓	↑ ↓	↑ ↓	↑	↑	O

electron configurations

C: $1s^2\ 2s^2\ p^2$
K: $1s^2\ 2s^2\ p^6\ 3s^2\ p^6\ 4s^1$
Cr: $3s^2\ p^6\ d^5\ 4s^1$

formulae for electron configuration

↑ ↓ denotes electron pair

electron configuration a description of the electrons and the orbitals they occupy in an atom. A closed shell is one in which all energy levels of the shell are occupied. The valence shell is an incomplete shell with some orbitals unoccupied; it is also called the outer shell to distinguish it from the inner, closed shells. In a neutral atom in the ground state, all electrons are in the lowest possible energy levels for stability reasons.

nuclear transition an energy change in the nucleus of an atom; it arises from the loss of an alpha or a beta particle (p.54) or the emission of a gamma ray (p.54).

electron transition the transference of an electron from one energy level to another in an atom. If energy is absorbed, the electron goes to a higher energy level. If the electron falls back to a lower energy level, radiation is emitted. The transitions of extra-nuclear electrons from higher to lower levels form the emission line spectra of atoms in the ultraviolet, visible and near infrared regions. Empty energy levels always exist in an atom above the energy levels occupied in the ground state.

atomic spectra an atom produces line spectra in various regions of the e.m. spectrum, depending on the method of excitation; each spectrum forms a series of lines. The simplest spectra are those of hydrogen. Providing sufficient energy is used in excitation, all spectra are produced at the same time. Each spectral series is produced by transitions of electrons to lower energy levels.

Lyman series a line spectrum of hydrogen in the ultraviolet region of the e.m. spectrum. The wave numbers of the lines are in agreement with the formula:

$$\sigma = R\ \frac{1}{1} - \frac{1}{m^2}$$

where R is the Rydberg constant (p.84), and $m > 1$. The ionization energy is determined by putting $m = \infty$,

electron transitions between extranuclear shells

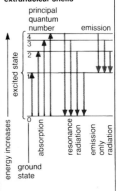

origins of spectral series for hydrogen

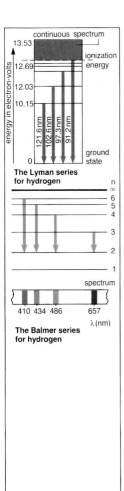

The Lyman series
for hydrogen

The Balmer series
for hydrogen

hence $\sigma = R = 10\ 973\ 731\,m^{-1}$, which corresponds to 13.53 eV. The electrons are raised from the ground state and eventually removed completely in ionization. The first term of the Rydberg equation has $n = 1$, for emission of radiation when the electrons fall back to the K-shell (p.45). Other spectra, particularly those of the alkali metals, are similar to the Lyman spectrum, but the lines are more complex.

Balmer series a line spectrum of hydrogen in the visible spectrum. The lines are in agreement with the formula

$$\sigma = R\left(\frac{1}{2^2} - \frac{1}{m^2}\right),$$

where $m > 2$. The series converges to a limit of $\sigma = R/4 = 2\ 743\ 433\,m^{-1}$. This represents the ionization energy for the first excited state of the hydrogen atom, when the electron has been raised to the L-shell, i.e. $n = 2$.

Paschen series a line spectrum of hydrogen in the near infrared region of the e.m. spectrum. The series converges to a limit of $\sigma = R/9$. This is the ionization energy from the M-shell, i.e. $n = 3$.

Brackett series a line spectrum of hydrogen in the infrared region of the e.m. spectrum. The series converges to a limit of $\sigma = R/16$.

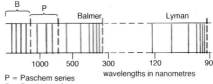

P = Paschem series
B = Brackett series
wavelengths in nanometres

Pfund series a line spectrum of hydrogen in the far infrared region of the e.m. spectrum. The series converges to a limit of $\sigma = R/25$.

fine structure the lines in a spectrum when examined under a more powerful spectrometer are often found to consist of two or more closely grouped lines. This is called the fine structure of a line.

hyperfine structure fine structure (↑) lines can sometimes be seen to consist of finer lines, this is the hyperfine structure.

multiplet (n) a group of lines in a fine structure. If it has two lines, it is called a **doublet** (p.90); with three lines a **triplet**.

doublet (*n*) a multiplet (p.89) with two lines. The lines arise from a difference in the spin quantum number, e.g. in the spectrum of sodium there is an intense yellow line of wavelength 589 nm. This arises from a transition of an electron from the 3 p to the 3 s energy level. In the transition, the electron can change its spin, then the energy of radiation is slightly different. Two lines appear in the doublet, and the difference of 0.6 nm in wavelength is due to the difference in spin energies.

triplet (*n*) *see* **multiplet**.

multiplicity (*n*) a measure of the number of lines in a fine structure (p.89).

fluorescence (*n*) the radiation of light from a molecule while it is excited by e.m. radiation in the visible or u.v. region. The emitted radiation is of longer wavelength than that of the radiation of excitation. When the molecule is excited, an electron is raised to a higher energy level and the spin can be changed. This forms a triplet energy level. If the multiplicity (↑) of both energy levels is the same, then the molecule fluoresces, and emits light of a longer wavelength, e.g. quinine sulphate under u.v. light fluoresces blue. **fluoresce** (*v*), **fluorescent** (*adj*).

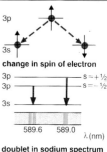

change in spin of electron

doublet in sodium spectrum

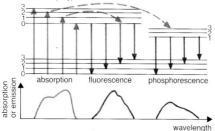

phosphorescence (*n*) the radiation of light from a molecule which continues after it has been excited by e.m. radiation in the visible or u.v. spectrum. The length of time of phosphorescence varies with different compounds. The emitted wavelength is longer than the absorbed wavelength of radiation. When a molecule is excited, an electron is raised to a higher energy level and the spin is changed. The multiplicity (p.90) of both levels is different, and the energy levels of the excited level are lower than that of the singlet state, i.e. the level without multiplicity. Calcium sulphide and zinc sulphide are phosphorescent substances. **phosphoresce** (*v*).

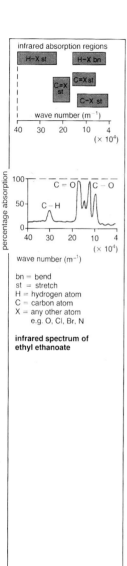

infrared absorption regions

wave number (m⁻¹)

bn = bend
st = stretch
H = hydrogen atom
C = carbon atom
X = any other atom
 e.g. O, Cl, Br, N

**infrared spectrum of
ethyl ethanoate**

infrared spectroscopy the spectrometer uses a heated rod of metal oxides as the source of illumination, a rock salt prism for forming the spectrum, a sensitive thermocouple to detect the intensity of the radiation for any given wavelength and a record of intensity is made by a pen on paper. Samples for examination are gases, liquids, or solutions. An absorption spectrum is recorded. If the frequency of the radiation source is the same as the frequency of the molecular vibration (p.92), then absorption occurs. The molecule must possess a dipole moment (p.75) across a bond for absorption to occur. Specific frequencies refer to specific vibrations and hence the structure of a molecule can be determined; *see diagram* for types of bonds detected by infrared spectrometers. In addition, for diatomic molecules, force constants for a bond can be calculated and hence the bond energy.
infrared spectrum.

Raman spectroscopy this method uses *scattered* light in the visible region of the e.m. spectrum. Monochromatic (p.92) light is used as a source and the scattered light examined by a spectrometer. A Raman spectrum is produced; this type of spectrum can show up molecular vibrations (p.92) which are inactive in the infrared spectra (↑). Most of the scattered light shows no change of frequency, but about 0.1 to 1% shows a slight change in frequency; this is **Raman scattering**. The slight change in frequency, Δν, is characteristic of a molecular species, as is the frequency in infrared spectra. No change in frequency produces a **Rayleigh line**; ν + Δν produces a **Stokes line**; ν − Δν produces an **anti-Stokes line**. The value of Δν is independent of the frequency of the source, and is a characteristic of a bond vibration.

Raman spectroscopy

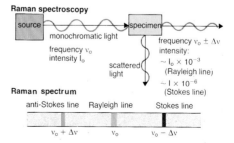

molecular vibration vibrational energy is quantized
(p.85). A molecule with two bonds can vibrate in three
ways, and these are shown, for linear and non-linear
molecules, *see diagram*. Stretching changes the value
of the dipole moment (p.75), as does bending. Non-
linear polyatomic molecules with x atoms can vibrate in
$(3x - 6)$ ways and linear molecules in $(3x - 5)$ ways. As
x increases, the spectra become very complex.

monochromatic (*adj*) describes a radiation which
consists of one wavelength only.

ultraviolet spectroscopy the spectrometer uses a
tungsten filament lamp as a source, a quartz prism to
form the spectrum, and a photocell to detect the
intensity of the radiation. A record of an absorption
spectrum is made and recorded by a pen on paper.
Specimens are usually dilute solutions or gases.
Ultraviolet spectroscopy is used: (1) with line spectra
(p.84) of atoms; (2) to find the structure of organic
compounds with double bonds; (3) in purity control of
pharmaceutical chemicals.

X-ray (*n*) X-radiation is produced when fast-moving
electrons from a cathode (cathode rays) strike a metal
target. The apparatus to produce X-radiation is shown
in *the diagram*. The origin of X-radiation is inelastic
collisions between the electrons and atoms in the
target. Energy is absorbed by the atom which produces
heat, and less than 1% of the energy is emitted by the
atom as an X-ray.

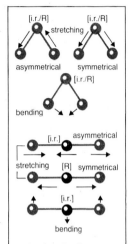

molecular vibration
i.r. measured in infrared
spectra

R measured in Raman
spectra

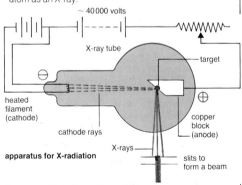

apparatus for X-radiation

X-ray spectrum the collisions between fast-moving
electrons and atoms of a target produce X-rays with a
continuous variation in frequency; this is called 'white'

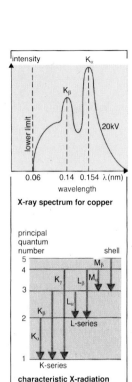

X-ray spectrum for copper

characteristic X-radiation

background emission. This background emission has a **cut-off wavelength**, or lower limit which depends only on the voltage between the anode and the cathode in an X-ray tube. Now $\nu = c/\lambda$ and $E = h\nu$, from Planck's constant (p.85) for the energy of the radiation. The energy of an electron is eV (measured in electron-volts), hence $eV = h\nu$, so for a tube voltage of 20kV,

$$\lambda = \frac{hc}{eV} = \frac{6.62 \times 10^{-34} \times 3 \times 10^8}{1.60 \times 10^{-19} \times 2 \times 10^4} \approx 6 \times 10^{-11}\,\text{m}$$

= 0.06nm. This agrees with the result from readings for a copper target. There are also peaks in the intensity of radiation, as well as background frequencies.

K-series (n) when a fast-moving electron, or an X-ray, hits an atom it can, if it has sufficient energy, knock an electron out of an inner orbital. This creates a hole in the orbital. An electron from an outer shell can then fall back into the hole, and emit an intense X-radiation, causing a peak in the spectrum. The electron can come from an L, M, N or higher shell and fall into the K-shell (p.45). The line spectrum formed by this emission is called the K-series. A line for the transition $L \rightarrow K$ is labelled α, from $M \rightarrow K$ is labelled β, and so on. Sub-levels of the K-shell, and other shells, form a fine structure on the peak. Each K-series is characteristic of an element.

L-series (n) this X-ray spectrum is similar to the K-series, but arises from electrons falling back into the L-shell (p.45). The lines are labelled L_α ($M \rightarrow$ L-shell), L_β etc. There are also M- and N-series.

Bragg's law the interatomic distances of crystals are of the same order of magnitude as the wavelengths of X-rays. A crystal lattice (p.165) will thus diffract X-rays. A simple explanation of diffraction is given in *the diagram*. When the distance (AB + BC) is equal to a multiple of the wavelength, interference (p.83) will produce a more intense beam, as the two wave crests will be additive. If the separation of layers in a crystal lattice is d and if θ is the angle at which the X-rays *glance* on the crystal, then (AB + BC) = $2d\sin\theta$. Other diffractions, called overtones, will take place for multiples of the wavelengths; let p be a multiple, then $p\lambda = 2d\sin\theta$, which is Bragg's law. The law can be used to establish crystal structures if X-ray wavelengths are known, or can be used to measure X-ray wavelengths if the crystal structure is known.

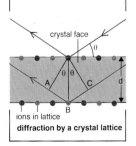

diffraction by a crystal lattice

Laue pattern the original method of Laue passed a beam of monochromatic X-rays through a crystal cut in a particular direction. This produced a pattern of spots on a flat photographic plate. The present method puts powdered crystals in the centre of a circle of photographic film, and the diffracted X-rays produce lines on the film. In each case, Bragg's law (p.93) is used to calculate the interatomic distances of the crystal lattice.

nuclear magnetic resonance the phenomenon of excitation of spin states in atomic nuclei, and measurement of the result. Abbreviation is nmr.

nmr spectrum a record of absorption as the applied magnetic field is decreased in strength by the sweep coil. The decreasing field strength is measured in parts per million (ppm) and can be related to resonance frequency in hertz.

electron spin resonance a magnetic field is produced by the spinning of an unpaired electron, i.e. one electron in an orbital. The spin can orient itself with or against an external magnetic field. This corresponds to two energy level states, and in an applied electric field an electron can flip from the lower to the higher energy state when it resonates to the correct frequency.

diamagnetism (n) the property of a substance or material which has a magnetic permeability less than unity. All substances and materials have this property which is due to the change in orbital motion of electrons in the presence of an applied external magnetic field. Diamagnetism is often overcome by paramagnetism (↓) or ferromagnetism (↓). **diamagnetic** (adj).

paramagnetism (n) the property of a substance or material which has a magnetic permeability slightly greater than unity. It is associated with unpaired electron spins in the orbitals of an atom. A paramagnetic substance tends to align the magnetic axes of its atoms with an applied external magnetic field. The permanent magnetic moment of an atom or ion (expressed in Bohr magnetons) = $\sqrt{[n(n+2)]}$ where n is the number of unpaired electrons in the atom or ion.

ferromagnetism (n) a property of iron, cobalt, nickel, and certain alloys which have an extremely high magnetic permeability. It is associated with a maximum number of unpaired electron spins in the orbitals of the atoms. Permanent magnets are made from ferromagnetic metals and certain alloys. **ferromagnetic** (adj).

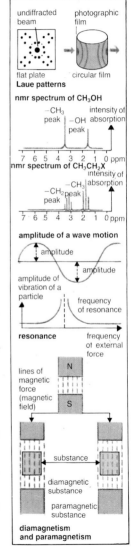

Laue patterns

nmr spectrum of CH_3OH

nmr spectrum of CH_3CH_2X

amplitude of a wave motion

resonance

diamagnetism and paramagnetism

Lambert-Beer law

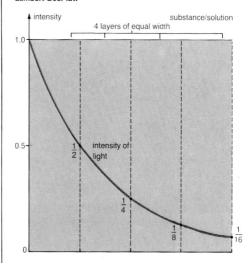

Lambert-Beer law the intensity of a beam of
monochromatic light is reduced by the same fraction in
each successive layer of equal thickness of a
homogenous material, e.g. if the intensity is reduced by
$\frac{1}{3}$ in the first 1 cm of the material, it will be further
reduced by $\frac{1}{3}$ in the next 1 cm and so on; thus after
passing through 1 cm, 2 cm and 3 cm of the material,
the intensity will be reduced to $\frac{1}{3}$, $\frac{1}{9}$, $\frac{1}{27}$ respectively of
the original intensity. The law is expressed
mathematically as: $I = I_o e^{-\alpha l}$, where I_o = incident (p.96)
intensity; I = intensity after a distance l through the
material and α is a constant for the material at a
particular frequency; α can also be written as α_ρ where
ρ is the frequency of the light. The law can also be
expressed as $\ln(I_o/I) = \alpha \, l$ or 2.303 log $(I_o/I) = \alpha l$. This
law is true for all substances and for all regions of the
e.m. spectrum. For a solution, absorption also depends
upon its concentration, and the law is restated as: log
$(I_o/I) = \varepsilon \, c$ / ϑηερε $\varepsilon = \alpha/2.303$ and c, the concentration,
is measured in moles per metre^3.
Beer-Lambert law another name for Lambert-Beer
law (↑).
Beer's law another name for Lambert-Beer law.

incident (*adj*) describes radiation or a moving body arriving at, or falling on, a surface. **incidence** (*n*).

optical density a measure of the ability of a material to absorb radiation; it has the symbol D.

$$D = \log \left(\frac{\text{incident intensity}}{\text{transmitted intensity}} \right) = \log \frac{I_o}{I}$$

Hence $D = \varepsilon c l$ (from the Lambert-Beer law (p.95). D has no units.

absorbance (*n*) another name for optical density (↑); it has the symbol A. $A = \varepsilon \ c \ l$. Absorbance is now preferred to optical density.

molecular extinction coefficient
(= ε expressed as ε metre^2 per mol)

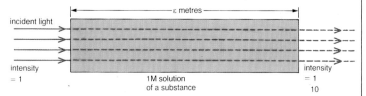

incident light

ε metres

intensity = 1

1M solution of a substance

intensity = $\frac{1}{10}$

molar extinction coefficient the thickness, in metres, of a molar solution which reduces the intensity of incident (↑) light to 1/10. It has the symbol ε. $\varepsilon = A/cl$, where A is the absorbance (↑). Units are m^2 mol^{-1}. ε is always determined experimentally; it is a property of a substance. The value of the coefficient varies with the frequency of the radiation and the temperature.

extinction coefficient the thickness, in metres, of a solution which reduces the intensity of incident (↑) light to 1/10. It has the symbol κ; $\kappa = \varepsilon c = A/\lambda$, where ε is the molar extinction coefficient (↑). Its units are m^{-1}.

dichroic (*adj*) describes crystals which have different colours when observed from different directions. A dichroic crystal absorbs light in one plane more than in a plane at right angles; if the crystal is sufficiently thick, it will absorb sufficient light in one plane to produce plane polarized (↓) light in the plane at right angles. Certain salts of quinine are dichroic, and when oriented correctly on a transparent medium produce plane polarized light, e.g. 'Polaroid' spectacles use such salts.

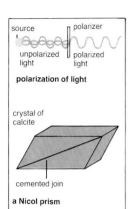

polarization of light

a Nicol prism

plane polarized describes an electromagnetic wave, e.g. light, in which the electric vector is in one plane only, called the plane of vibration; the corresponding magnetic vector is in a plane at right angles to the plane of vibration, this is called the plane of polarization. The waves of an e.m. radiation are emitted from a source with the directions of the vibrations at any angle in a plane perpendicular to the line of transmission. When plane polarized the vibrations are confined to one plane.

Nicol prism a crystal of calcite is cut into two pieces with precise angles and the two halves are cemented together with Canada balsam. This forms a Nicol prism and it causes light to be plane polarized (↑).

optical activity a property possessed by certain substances (mainly organic) and their solutions, of rotating the plane of polarized (↑) light. The degree of rotation in a solution is proportional to the concentration of the solute, the distance the light travels through the solution, and the wavelength of the light.

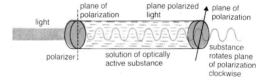

optical activity

rotary dispersion the degree of rotation caused by an optically active (↑) substance in solution depends on the wavelength of the plane polarized (↑) light. White light is *dispersed* into the colours of the spectrum by an optically active substance.

polarizer (*n*) any crystal, or a Nicol prism (↑), which produces plane polarized (↑) light.

analyzer (*n*) a polarizer (↑) used to detect the plane of polarization. When the planes of polarization in a polarizer and an analyzer are at right angles, no light passes through. The polarizer and analyzer are then said to be 'crossed'.

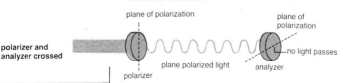

polarizer and analyzer crossed

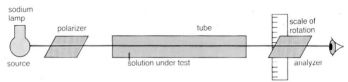

diagram of a polarimeter

polarimeter (*n*) an instrument for measuring optical activity (p.97). It consists of a tube for containing solutions, two Nicol prisms, a polarizer and an analyzer (p.95), and an eyepiece (a lens to magnify the intensity of the light). A monochromatic light source is used, this is usually a sodium lamp. The analyzer is mounted on a scale to measure the angle of rotation. The degree of rotation is found by turning the analyzer until it is crossed with the polarizer. **polarimetry** (*n*).

dextrorotatory (*adj*) describes optical activity (p.97) which rotates the plane of polarization of light clockwise when viewed in the direction of the light source. Such rotation is labelled +, i.e. positive.

laevorotatory (*adj*) describes optical activity (p.97) which rotates the plane of polarization of light anticlockwise when viewed in the direction of the light source. Such rotation is labelled −, i.e. negative.

racemic (*adj*) describes a mixture of equal proportions of dextrorotatory (↑) and laevorotatory (↑) forms of an optically active compound. *See* **optical isomerism** (p.32).

resolution (*n*) the separation of a racemic (↑) mixture into separate dextrorotatory (↑) and laevorotatory (↑) isomers (p.31).

meso- (pre) if a compound has two asymmetric carbon atoms, then the dextrorotatory (↑) and the laevorotatory (↑) activities can be balanced and the compound is not optically active. Such a compound is 2, 3,–dihydroxybutanedioic acid (tartaric acid). The acid exists in +, −, racemic and meso-forms.

racemization (*n*) the changing of an optically active (p.97) form of a substance, either + or −, into a racemic (↑) mixture. This may be done by boiling or by a chemical action. If the change takes place by itself it is called **autoracemization**. *See* **racemate** (p.32).

optical rotation (1) another name for **optical activity** (p.97). (2) the angle of rotation of plane polarized light by an optically active compound.

dextrorotatory optical activity

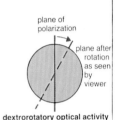

laevorotatory optical activity

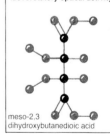

meso-2,3 dihydroxybutanedioic acid

COOH		COOH
H–C–OH		HO–C–H
HO–C–H		H–C–OH
COOH		COOH

optical isomers of 2,3,dihydroxybutanedioic acid: (+) and (−) forms

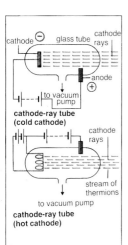

cathode-ray tube
(cold cathode)

cathode-ray tube
(hot cathode)

a simple electron gun

cathode-ray tube a glass tube evacuated to a low pressure (less than 0.01 mm of mercury) with a cathode and an anode sealed in the glass walls. A high voltage is applied between the anode and cathode, and a green fluorescence appears at the end of the tube directly opposite the cathode. The fluorescence is caused by bombardment with cathode rays.

cold cathode a metal plate used as a cathode in a cathode-ray tube (↑).

hot cathode a hot wire filament used as the cathode in a cathode-ray tube. The filament is heated by a small battery. The voltage required to work the tube with a hot cathode is much lower than is used with a cold cathode.

cathode rays rays produced by a cathode-ray tube (↑). The rays can be deflected by magnetic and electrical fields. They consist of a beam of fast-moving electrons, whose speed depends on the voltage applied between the cathode and the anode. The material of the cathode has no effect on the production of cathode rays; this is evidence that all matter contains electrons.

discharge tube a glass tube, containing gas at a low pressure, with a cathode and an anode. When a voltage is applied to the electrodes, the gas glows and emits light which is the visible spectrum (p.82) for the particular gas.

luminescence (n) (1) any process which includes the emission of light. (2) the light emitted by such a process, e.g. the luminescence caused by cathode rays (↑). **luminescent** (adj).

electron gun a glass vessel, evacuated to a low pressure, containing a hot cathode (↑) and an anode. The cathode is surrounded by a metal cylinder which ends in a wire mesh. The cylinder is kept at a negative potential to the cathode so that it concentrates the electrons emitted by the cathode. The anode attracts the electrons, and they pass through a small hole in the anode forming a narrow beam of electrons.

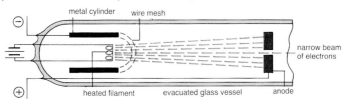

positive rays a discharge tube (p.97) with a perforated cathode, and a plate anode is filled with a gas at a low pressure. When a high voltage is applied between the anode and cathode, positive rays are emitted from the cathode and travel in a direction away from both the anode and the cathode. The positive rays are deflected by magnetic and electrical fields and are shown to have a positive charge. They are positive ions (p.124) from the gas in the tube.

photoelectric effect the emission of electrons from the surface of a solid when an electromagnetic radiation of sufficiently high frequency is incident (p.96) on the surface. With the alkali metals, e.m. radiation in the visible region causes emission; for other metals, u.v. light is required to cause emission. The following conditions are observed: (1) there is no emission below a certain frequency of radiation, called the **threshold frequency**; (2) for a given frequency, greater than the threshold frequency, the kinetic energy of the emitted electrons has a range of values from zero up to a maximum. The maximum energy is proportional to $(\nu - \nu_o)$, where ν is the incident frequency and ν_o is the threshold frequency; (3) emission occurs, with suitable frequencies of radiation, however weak is the intensity (p.83) of radiation; (4) the number of electrons emitted, i.e. an electric current, is proportional to the intensity of the radiation; (5) the threshold frequency is a property of the material.

electron diffraction if a beam of electrons has the nature of a wave motion, then de Broglie's equation (p.85) shows that the wavelengths of the motion are of the same order of magnitude as the interatomic distances between atoms in metal crystals. The

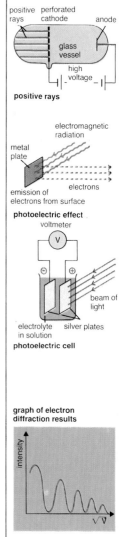

positive rays

photoelectric effect

emission of electrons from surface

photoelectric cell

graph of electron diffraction results

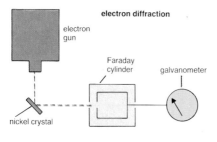

electron diffraction

electron gun

Faraday cylinder

galvanometer

nickel crystal

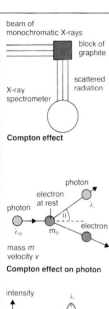

Compton effect

Compton effect on photon

mass m
velocity v

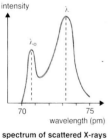

spectrum of scattered X-rays

apparatus to investigate electron diffraction is shown in *the diagram*. A beam of electrons from an electron gun (p.99) hits the surface of a crystal of nickel. The resulting radiation is collected in a Faraday cylinder, and a galvanometer records the intensity of the radiation. Altering the applied potential V of the electron gun alters the energy of the electrons. Energy of electrons = $eV = \frac{1}{2}mv^2$, where e = charge on electron, V = potential applied in electron gun, m = mass of electron and v = its velocity. Therefore $v \propto \sqrt{V}$, but $mv = \frac{h}{\lambda}$ from de Broglie's equation, therefore $\frac{1}{\lambda} \propto \sqrt{V}$. Bragg's Law (p.93) states intensity $\propto \frac{2d \sin \theta}{n\lambda}$, i.e. intensity $\propto 1/\lambda$ and hence $\propto \sqrt{V}$. The peaks in the graph correspond to the integral diffraction bands in X-ray diffraction. Hence a beam of electrons is diffracted and hence has a wave nature.

Compton effect matter scatters electromagnetic radiation. For wavelengths of visible light, and longer wavelengths, the scattered beam has the same wavelength as the incident (p.96) beam. For shorter wavelengths, i.e. u.v. light, X-rays and gamma rays, a second radiation of longer wavelength appears in the scattered radiation. This is the Compton effect and it is due to collisions between photons and free electrons. Part of the photon's energy is transferred to the electron, so the photon, after collision, has a lower energy, and hence a longer wavelength.

mass-energy equation matter and energy can be converted from one to the other under suitable condit-ions. Einstein's equation for the conversion is $E = mc^2$, where a mass, m, is converted to energy E, c is the vel-ocity of light. If 1 g of matter is converted to energy, then

$$E = 0.001 \, \text{kg} \times (3 \times 10^8)^2 = 9 \times 10^{13} \, \text{J} = 9 \times 10^{10} \, \text{kJ}.$$

dual nature of matter in experimental investigations, results indicate that matter is composed of atoms and sub-atomic particles, such as the electron. The particulate (p.100) nature of matter is studied in particle mechanics and in the quantum theory (p.85), where both mass and energy are considered to be particulate. In other experimental investigations, such as electron diffraction, results indicate that matter has a wave nature. Matter thus has a dual nature; it can behave as a set of particles or as a set of waves.

uncertainty principle a principle, stated by Heisenberg, that it is impossible to determine precisely the position and the momentum of a particle. The more precisely one is known, the less accurately is the other known, according to the equation $\Delta x.\Delta p = h$ where Δx is the range in position, Δp is the range in momentum and h is Planck's constant (p.85). If a particle of mass m is travelling with a velocity v, then $\Delta x.\Delta v = h/m$, indicating that the uncertainty principle is only of importance for sub-atomic particles. A similar statement exists for energy, E, and time, t, i.e. $\Delta E.\Delta t = h$. For electrons in atoms, only the probability of location can be given accurately.

particulate (*adj*) consisting of particles.

momentum (*n*) the product of mass and velocity of a body.

mass defect the mass of a nucleus is always slightly different from the sum of the masses of protons and neutrons. The difference is the mass defect. If a nucleus consists of Z protons and N neutrons, and the mass of each is m_p and m_n respectively, then the mass defect = $Zm_p + Nm_n - m^*$ where m^* is the measured mass of the nucleus. Mass defect has the symbol m_d.

binding energy (1) the work done to separate the components of a nucleus into separate protons and neutrons. Conversely, it is the energy emitted when the nucleus is built up from its constituent protons and neutrons. If m_d is the mass defect (↑), then the binding energy, B, is given by $B = m_d c^2$, using the mass-energy equation (p.101). For energy conversions, 1 atomic mass unit = 931 meV. *See diagram* for the graph of binding energy against mass number. (2) the binding energy of one proton or one neutron to a nucleus.

packing fraction this is defined from:

$$\text{packing fraction} = \frac{\text{isotope mass} - \text{mass number}}{\text{mass number}}$$

In the case of $^{12}_6\text{C}$ the packing fraction is zero because the isotope mass is taken as the mass number, e.g. the packing fraction for $^{35}_{17}\text{Cl}$ is:

$$\frac{34.96885}{35} = -8.9 \times 10^{-4}$$

The lower is the packing fraction, the greater is the binding energy (↑) per nucleon (↓), so the curves for packing fraction and binding energy are of the same shape, but the reverse of each other.

nucleon (*n*) one of the particles which gives a nucleus (p.45) its mass, i.e. a proton or a neutron.

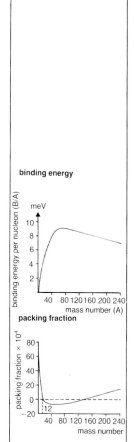

binding energy

binding energy per nucleon (B/A)

meV

10
8
6
4
2

40 80 120 160 200 240
mass number (A)

packing fraction

packing fraction × 10⁴

80
60
40
20
0
-20

12

40 80 120 160 200 240
mass number

standard state the standard state of a compound or element is defined as its most stable physical form at a pressure of 1 atmosphere and a specified temperature, usually 298 K (25°C). By convention, an element in its standard state is assigned an enthalpy (p.112) of zero.

heat change heat is a method of transferring energy because of a temperature difference. The symbol for heat is q. Heat added to a chemical system appears as a change in the internal energy (p.111) of the system or as work, w, done by the system. Expressed mathematically, $q = \Delta U + w$. Heat changes are measured at constant pressure, preferably, or constant volume, as under these simple conditions:
$q = \Delta U$ (change in internal energy) at constant volume
$q = \Delta H$ (change in enthalpy) at constant pressure. At constant volume, no external work is performed. At constant pressure, $\Delta H = \Delta U + w$.

the heat of reaction for the combustion of benzene

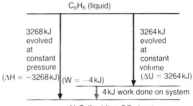

C_6H_6 (liquid)

3268 kJ evolved at constant pressure ($\Delta H = -3268$ kJ)

$(W = -4$ kJ$)$

3264 kJ evolved at constant volume ($\Delta U = 3264$ kJ)

4 kJ work done on system

H_2O (liquid) $+$ CO_2 (gas)

heat of reaction the heat given out, or absorbed, during a chemical reaction. As the heat of reaction depends on the amounts of the reactants, the heat evolved, or absorbed, is usually determined for one mole of one of the reactants, e.g. in the equation:

$C_6H_6(l) + 7\frac{1}{2}O_2(g) = 3H_2O(l) + 6CO_2(g) + 3268$ kJ

it is assumed that one mole of liquid benzene is completely burnt in oxygen at atmospheric pressure to form liquid water and carbon dioxide gas, with the products ending in their standard states (↑). The equation shows that 3268 kJ of heat are given out. The convention for internal energy and for enthalpy uses a negative sign for heat given out, as the system has lost heat.

heat of combustion the heat change when one mole of a substance is completely burnt in oxygen. The products of combustion of organic compounds are gaseous carbon dioxide and liquid water. If the substances are in their standard states, the result is the standard heat of combustion.

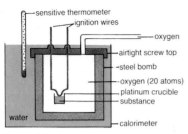

bomb calorimeter

bomb calorimeter a device for measuring heats of combustion. An organic substance is placed in a platinum crucible enclosed in a thick-walled steel bomb which has a screw top. An oxygen supply, delivered through a valve system, fills the bomb with oxygen at 20 atmospheres pressure. The substance is ignited by a thin wire passing an electric current. The bomb is placed in a calorimeter containing water. The heat of the reaction is measured from the temperature rise of the water in the calorimeter; the heat measurement is at constant volume.

heat of formation the heat change when 1 mole of a substance is formed from its elements. To achieve standardization, standard states of elements and compounds are used for the measurement.

enthalpy of formation the heat of formation at constant pressure is the enthalpy of formation. Most enthalpies of formation are measured indirectly, e.g. the enthalpy of methane is measured by:

$CH_4(g) + 2O_2(g) \rightarrow CO_2(g) + 2H_2O(g)$ $\Delta H = -802$ kJ
$C(s) + O_2(g) \rightarrow CO_2(g)$ $\Delta H = -393$ kJ
$2H_2(g) + O_2(g) \rightarrow 2H_2O(g)$ $\Delta H = 2(-242)$ kJ
$C(s) + 2H_2(g) \rightarrow CH_4(g)$ $\Delta H = 2(-242) - 393 + 802$
$= -75$ kJ

ΔH is the enthalpy change for each of the three combustion reactions. The state of each reactant and product is shown: (g) for gas; (s) for solid. Tables of standard enthalpies of formation are available.

heat of atomization the heat change when one mole of an element in its standard state of 298 K and one atmosphere pressure is converted into free atoms. The heat of atomization is usually obtained from spectroscopic measurements. Heats of atomization are used to measure bond energies (p.73).

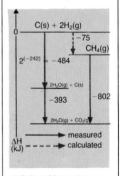

enthalpy of formation of methane

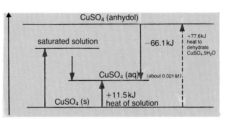

heats of solution and dilution

heat of solution the heat change when 1 mole of a substance dissolves in such a large volume of solvent that addition of more solvent produces no further change. The subscript (aq) is used to show infinite dilution, the dilution at which the heat of solution is measured. *See* **heat of dilution** ($\downarrow$).

heats of dilution for $CuCl_2,2H_2O$

molality (m) mol	2.8	1.2	0.6	0.3
heat change (ΔH) kJ	-49	-44	-26	-16
integral heat of solution ($\Delta H/m$) kJ/mol	-17.5	-37	-47	-50
heat of dilution		19.5	10	3

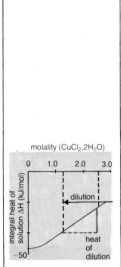

molality ($CuCl_2,2H_2O$)

heat of dilution

heat of dilution the heat evolved or absorbed when a solution is diluted. The integral heat of dilution is the heat change when a solution containing one mole of solute is diluted from one concentration to another. It can be seen from the table for crystalline copper (II) chloride, i.e. $CuCl_2, 2H_2O$, that the heat of dilution approaches zero as the solution becomes very dilute, i.e. further dilution produces no heat change; at infinite dilution the integral heat of solution becomes the standard heat of solution. Thermochemical equations for heats of dilution are expressed as:

$$HCl (50 H_2O) + aq \rightarrow HCl (aq) \ \Delta H = -1.84 kJ.$$

enthalpy of reaction the heat of reaction (p.103) measured at constant pressure. The enthalpy of any reaction can be calculated from the difference between the enthalpies of all the products and the enthalpies of all the reactants. If ΔH is the enthalpy of reaction, ΔH_f is the enthalpy of formation, and Σ stands for 'the sum of', then: $\Delta H = \Sigma \Delta H_f \text{(products)} - \Sigma \Delta H_f \text{(reactants)}$.

internal energy of reaction the heat of reaction (p.103) measured at constant volume. For reactions involving only solids and/or liquids, volume changes are negligible, so the internal energy of a reaction is approximately equal to the enthalpy of reaction. For reactions involving gases, an expansion or a contraction of volume may be expected. For constant volume measurements, this appears as heat. The relationship between internal energy changes, ΔU, and enthalpy changes, ΔH, is given by: $\Delta U = \Delta H \pm nRT$, where n is the mole fraction change for gases, R is the molar gas constant, and T is the thermodynamic temperature (p.108). If there is an increase in gas volume, the negative sign is used. Values of ΔU for a reaction can be calculated from the difference between the sum of the internal energies of products and reactants as for enthalpies of reaction (p.103).

internal energy of formation the heat of formation (p.104) at constant volume.

Hess's law the heat evolved or absorbed in a chemical change either at constant pressure or constant volume is the same whether the change is brought about in one stage or several stages, e.g. the formation of carbon dioxide from carbon and oxygen can follow two paths. The enthalpy of reaction (p.103) is the same for both paths. A number of reactions can be combined algebraically to calculate an enthalpy of reaction, e.g. the enthalpy of the reaction between carbon and steam can be calculated as follows:

$C(s) + \frac{1}{2}O_2(g) \rightarrow CO(g)$		$\Delta H = -111\,kJ$
$H_2O(g) \qquad\quad \rightarrow H_2(g) + \frac{1}{2}(O_2)g$		$\Delta H = +242\,kJ$

$$C(s) + H_2O(g) \rightarrow CO(g) + H_2(g) \qquad \Delta H = +131\,kJ$$

Note that the reaction for the formation of water has been reversed, *see energy diagram*; oxygen on each side of the reaction cancels out.

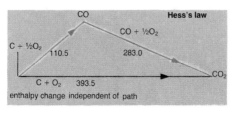

enthalpy change independent of path

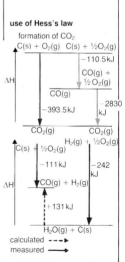

use of Hess's law

formation of CO_2

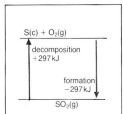

law of Lavoisier and Laplace

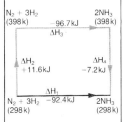

Kirchoff's law

law of constant heat summation *see* **Hess's law** (↑).

law of Lavoisier and Laplace the quantity of heat that must be supplied to decompose a compound into its elements is equal to the heat evolved when the compound is formed from its elements. The law can be extended to all types of reactions, e.g.

$C_2H_6(g) + 3\frac{1}{2}O_2(g) \rightarrow 2CO_2(g) + 2H_2O(l)$
$$\Delta H = -1560 \text{ kJ}$$
$2CO_2(g) + 2H_2O(l) \rightarrow C_2H_6(g) + 3\frac{1}{2}O_2(g)$
$$\Delta H = +1560 \text{ kJ}.$$

enthalpy of a compound by convention, the enthalpies of all elements in their standard states are arbitrarily taken to be zero, hence the enthalpy of a compound is equal to the enthalpy of formation (p.104).

Kirchhoff equation the heat of reaction varies with temperature, and Kirchhoff's equation gives the variation of enthalpy of reaction with temperature. The diagram shows the reaction $N_2 + 3H_2 \rightarrow 2NH_3$. ΔH_1 is the standard enthalpy of reaction, -92.4 kJ. ΔH_2 is calculated from the molar heat capacities at constant pressure. For N_2 this is 29.0 J/K; for H_2 this is 28.9 J/K, and for NH_3 this is 36.1 J/K. For $N_2 + 3H_2$, $\Delta H_2 = 29.0 + 3(28.9) = 115.6$ J/K; for a temperature difference of 100 K,

$$\Delta H_2 = 115.6 \times 100 \div 1000 = 11.6 \text{ kJ}.$$
$$\Delta H_4 = 2 \times 36.1 \text{ J/K} = 7.2 \text{ kJ for } 100 \text{ K}.$$

Now $\Delta H_1 = \Delta H_2 + \Delta H_3 + \Delta H_4$, hence $\Delta H_3 = -92.4 - 11.6 + 7.2 = 96.7$ kJ. Kirchhoff's law relates ΔH to C_p (molar heat capacity at constant pressure) by Kirchhoff's equation which is:

$$\frac{d}{dT}(\Delta H) = \Delta C_p.$$

C_p does change with temperature, and in the example, a constant value is taken. For small temperature differences, however, the error is only small, in this case 0.1%.

Gay-Lussac's law at constant pressure, the volume of a fixed mass of any gas increases by the same fraction for every degree rise in temperature. For a temperature difference, t, the volume increases from V_0 to V_t according to the equation: $V_t - V_0 = V_0 \alpha t$, where α is a constant for an ideal gas. Isobars (↓) for a fixed mass of gas are shown in the diagram.

isobar[1] (*n*) a graph showing the relationship between two variables at a constant pressure.

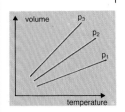

isobars for a fixed mass of an ideal gas

thermal expansivity if small changes are made in temperature, ΔT, corresponding small changes in volume, ΔV, of a gas are produced, so: $\Delta V = V\alpha\Delta T$, from Gay-Lussac's law (p.105). For infinitesimal changes, $\alpha = V^{-1}\,\delta V/\delta T$ at constant pressure for a fixed mass of gas; α is the thermal expansivity of the gas.

absolute temperature at low densities, all gases at constant volume and of fixed mass show the same fractional increase in pressure for every degree rise in temperature, as measured by a mercury thermometer. The relationship is: $P_t = P_o(1 + \alpha t)$ where t is the temperature rise and α is the thermal expansivity (↑). If the difference between the boiling point and the freezing point of water is made equal to 100°, the $\alpha = 1/273.15$. The ratio of two temperatures, t_1 and t_2, is defined for an ideal gas (p.147), by:

$$\frac{P_1}{P_2} = \frac{P_o(1/\alpha + t_1)}{P_o(1/\alpha + t_2)} = \frac{273.15 + t_1}{273.15 + t_2} = \frac{T_1}{T_2}$$

where $T = (273.15 + t)$. The temperature, T, is an absolute temperature determined from the measurements of pressure of a constant volume of an ideal gas.

thermodynamic temperature (1) absolute temperature (↑). (2) temperature defined from thermal efficiency; it has the same value as absolute temperature, but the definition is independent of any substance.

work (n) mechanical work, w, is determined by a force, f, moving its point of application a distance, l, in the direction of the force; the equation is:

$$w = \int_{l_1}^{l_2} f\,dl.$$

When an external force is applied to a piston of area A, operating on a gas in a cylinder, then:

$$w = \int_{l_1}^{l_2} \frac{f}{A}.A\,dl = \int_{V_1}^{V_2} p\,dV$$

since $V = Al$ and $f/A = p$, the external pressure. Work is defined by a process, and two different processes between identical initial and final states produce two different amounts of work. In the diagram, the work done by the system on its surroundings is greater for process 1 than for process 2.

thermodynamically reversible change a change in which a chemical system is always in temperature and

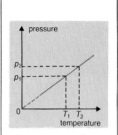

graph of pressure against temperature for a constant volume of a fixed mass of an ideal gas

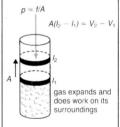

$p = f/A$

$A(l_2 - l_1) = V_2 - V_1$

gas expands and does work on its surroundings

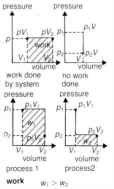

work done by system · no work done

process 1 · process2

work $\quad w_1 > w_2$

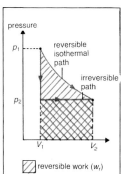

reversible work (w_r)

irreversible work (w_i)

$w_i = p_2 (V_2 - V_1)$

$w_r = RT \ln (V_2/V_1)$

**reversible and
irreversible changes**

pressure equilibrium with its surroundings. To do this, the change has to be carried out infinitesimally slowly, a process almost impossible to perform practically. In a reversible process, the entropy (p.113) change of the system equals the entropy change of the surroundings, i.e. $\Delta S = q/T$ and $q_{in} = q_{out}$ for a reversible process, where S is the entropy, and q the heat change. *See* **irreversible change** ($\downarrow$). There is no overall change in entropy as entropy gained by the system equals entropy lost by the surroundings.

irreversible change the change in entropy (p.113) of a chemical system depends only on the initial and final states and $\Delta S = q_{rev}/T$, where ΔS is the entropy change of the system, q_{rev} is the reversible heat change, and T is the absolute temperature. The heat change, q, of the system, however, depends on the path followed by the irreversible change. In an irreversible expansion, the external pressure is always lower than the internal pressure (to cause expansion), hence the irreversible work done by the system is lower than the reversible work. The change in internal energy, ΔU, is the same whichever path is followed, as it depends only on the initial and final states, so:

$$\Delta U = q_{rev} - w_{rev} = q_{irrev} - w_{irrev}$$
$$\text{therefore } q_{rev} - q_{irrev} = w_{rev} - w_{irrev}$$
$$\text{since } w_{rev} > w_{irrev} \text{ hence } q_{rev} > q_{irrev}$$

The heat absorbed or evolved in an isothermal reversible change is greater than the heat absorbed or evolved in an irreversible change. The total entropy change is given by:

$$\Delta S = \frac{q_{rev}}{T} - \frac{q_{irrrev}}{T} > 0$$

i.e. there is an increase in total entropy.

maximum work maximum work is obtained from a reversible change ($\uparrow$) as $w_{rev} > w_{irrev}$. The increase in entropy (p.113) of a chemical system for a reversible change is given by $\Delta S = q_{rev}/T$ at constant temperature. Now $\Delta A = \Delta U - T\Delta S$, where U is the internal energy (p.111) and A is the Helmholtz free energy (p.113). Now $q_{rev} = \Delta U + w_{rev}$, so $\Delta A = \Delta U - q_{rev}$, but $-w_{rev} = \Delta U - q_{rev}$ for the reversible change, so $\Delta A = -w_{rev}$.

maximum work function the property, A, is called the maximum work factor since $\Delta A = -w_{rev}$ = maximum work ($\uparrow$). It is also the Helmholtz free energy (p.113).

net work the change in the Gibbs free energy (p.112) is given by: $\Delta G = \Delta A + P\Delta V = -w_{rev} + P\Delta V$, i.e. $-\Delta G = w_{rev} - P\Delta V$ at constant temperature and pressure. $P\Delta V$ is the work done by expansion against the external pressure so $-\Delta G$ is the maximum work at constant temperature and pressure, other than that due to a volume change. The quantity $w_{rev} - P\Delta V$ is the *net work*, and so a decrease in ΔG is equal to the net work obtainable from a chemical system under reversible conditions at constant temperature and pressure.

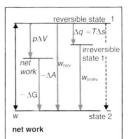

net work

spontaneous change a spontaneous change is a chemical process occurring without external aid; it is always thermodynamically irreversible, as it takes place at a finite rate. For an irreversible process at constant temperature and pressure, ΔG must be negative, i.e. the Gibbs free energy of the system diminishes. The standard Gibbs free energy of a reaction can be derived from the e.m.f. of a cell, by using the equation $\Delta G = -nFE^o$, where E^o is the standard e.m.f. of a reversible cell using the reaction. If the value of ΔG from this measurement is negative, then a spontaneous reaction is possible. In a *chemically* reversible reaction, the reaction takes place in such a direction as to diminish the free energy. When $\Delta G = 0$, the reaction is in equilibrium.

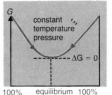

chemically reversible reaction

reversible cycle a cyclic process consisting of a succession of changes such that a system returns to its original state. The internal energy change, ΔU, is equal to zero. Maximum work is obtained by performing thermodynamically reversible changes (p.108) and the work is equal to the net heat absorbed. A reversible cycle is shown in the diagram. AB is an isothermal expansion of a gas at temperature T_1, work w_1 is done *on* the surroundings. BC takes place at constant volume, the gas is cooled and its pressure falls, no work is done. CD is an isothermal compression at temperature T_2, work w_2 is done *by* the surroundings. DA takes place at constant volume, the gas is heated and its pressure rises, no work is done. Now:
$w_1 = RT_1 \ln V_2/V_1$ $w_2 = RT_2 \ln V_1/V_2 = -RT_2 \ln V_2/V_1$
net work $= w_1 + w_2 = R(T_1 - T_2) \ln V_2/V_1$.

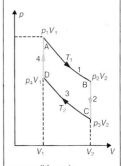

a reversible cycle

thermodynamic efficiency in a reversible cycle (↑) heat q_2 is absorbed at a higher temperature T_2, and heat q_1 is evolved at a lower temperature T_1. The efficiency of the cycle is the fraction of heat absorbed at the higher temperature that is converted to work. The total entropy

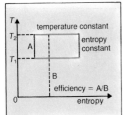

thermodynamic efficiency

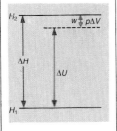

**enthalpy and
internal energy**

(p.113) change for the complete cycle is zero, as the gas returns to its original state hence:

$$\frac{q_2}{T_2} - \frac{q_1}{T_1} = 0,$$

$$\text{i.e. } \frac{q_2}{T_2} = \frac{q_1}{T_1}$$

$$\text{and efficiency} = \frac{q_2 - q_1}{q_2} = \frac{T_2 - T_1}{T_2}$$

first law of thermodynamics energy cannot be created or destroyed, but is always transformed to another form, and is thus conserved (work is done by the transformation).

second law of thermodynamics there are several statements of this law: (1) heat cannot be completely converted into an equivalent amount of work without causing changes in some part of a system, or of its surroundings. (2) a quantity, called entropy (p.113), is a state function of a system. In a reversible process, entropy remains constant, but in an irreversible process, entropy increases, both effects acting on the system and its surroundings. In no process does the entropy of the universe decrease.

third law of thermodynamics the entropy (p.113) of a perfect crystalline solid of a pure substance is zero at a temperature of absolute zero.

zeroth law of thermodynamics if systems A and B are separately in thermal equilibrium with system C, then systems A and B are in thermal equilibrium with each other (all systems in thermal equilibrium with one another are said to be at the same temperature).

internal energy the energy possessed by a chemical system. The internal energy is determined by the pressure, temperature and composition of the system together with the kinetic energy of individual molecules or ions and the potential energy of elementary particles within the individual molecules or ions; it is the total energy possessed by the system. Internal energy of a chemical system cannot be determined experimentally, but the *change* in internal energy can be determined. The symbol for internal energy is U, and for a change in internal energy is ΔU. Now $\Delta U = q - w$ where q is the heat energy taken from the surroundings, and w is the work done on the surroundings. Internal energy is an extensive property.(p.

intrinsic energy another name for **internal energy** (↑).

enthalpy (n) the heat energy content of a system. It is the heat energy given out, or absorbed during a chemical process. The **enthalpy of a compound** is equal to the heat given out or taken in when one mole of the compound is formed from its constituent elements (in their standard states) at constant pressure. Elements in their standard states are given zero enthalpy, by convention. The symbol for enthalpy is H, and for a change in enthalpy, ΔH. Heat given out is $-\Delta H$ and heat taken in is $+\Delta H$, in a chemical reaction, e.g.

$$C + \tfrac{1}{2}O_2 \rightarrow CO \; \Delta H = -110.4\,kJ.$$

The reaction is exothermic; $-110.4\,kJ$ is the enthalpy of carbon monoxide, with the value given for the formation of one mole. The relation between internal energy and enthalpy is $H = U + pV$, where p is pressure and V is volume. For a change in enthalpy, $\Delta H = \Delta U + p\Delta V$. For gases, $p\Delta V = w$, i.e. the work done on the surroundings. For solids and liquids, w is small, and $\Delta H \approx \Delta U$. ΔH is measured in heats of reaction (p.103).

heat content another name for **enthalpy** (↑).

chemical system a system of one, two or more phases (p.159) where each phase is homogenous (p.161) and may consist of one or more chemical substances.

Gibbs free energy the energy available in a chemical system to do work at a constant pressure; the symbol for this energy is G. The characteristic of G is that at constant pressure and temperature a process takes place only in the direction where G decreases. If ΔG is the change in G, then a process can occur if $\Delta G < 0$, but not if $\Delta G > 0$. In an equilibrium state, at constant pressure and temperature, G tends to a minimum value. Gibbs free energy is defined from the relation $G = H - TS$, where H is the enthalpy (↑), T the absolute temperature, and S the entropy (↓) of a chemical system. G is measured in joules per mole, it is an extensive property. G can be calculated from equilibrium constants (p.185) and standard electrode potentials (p.192). The relation is $\Delta G^\circ = -RT \ln K$, and $\Delta G^\circ = -nE^\circ F$ where R is the gas constant, E° is the standard electrode potential, n the number of electrons transferred in the reaction, F is the faraday (p.127), and ΔG° is the standard Gibbs free energy.

standard free energy the Gibbs free energy (↑), or a change in free energy, measured at 298K, and a pressure of 1 atmosphere.

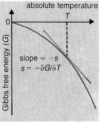

value of G and S at temperature T

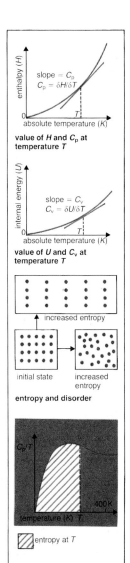

value of H and C_p at
temperature T

value of U and C_v at
temperature T

entropy and disorder

⬜ entropy at T

Helmholtz free energy the energy available in a chemical system to do work at a constant volume; the symbol for this energy is A. It is defined from the relation $A = U - TS$, where U is the internal energy (p.111), T the absolute temperature and S is the entropy (↓). This quantity is not used much in chemical calculations, as most experiments are carried out at constant pressure.

heat capacity the number of joules needed to raise the temperature of an object by 1 K (1°C). For pure substances the **molar heat capacity** is the heat capacity of one mole of substance, and is the quantity normally used. The molar heat capacity varies with temperature and enthalpy as shown in the graph. This is the heat capacity at constant pressure, the symbol is C_p. The heat capacity at constant volume, C_v, is determined from the graph of internal energy against temperature.

entropy (n) a measure of the amount of disorder in a system; the higher the entropy, the greater the disorder. Disorder can be molecular, i.e. when a solid changes to a liquid at the melting point, or the result of mixing two gases. Entropy changes when heat is taken in by a substance, and it is measured as the heat absorbed divided by the temperature at which it is absorbed. The symbol for entropy is S, so $S = q/T$, where q is the heat absorbed, and T is the absolute temperature. At the melting point of a solid, heat is taken in at a constant temperature, so the latent heat of fusion is the entropy of fusion. Thermal energy consists of two factors, a quality (temperature) and an intensity (entropy), this corresponds to frequency and amplitude of a wave. In reversible reactions (p.185) entropy is conserved, but in irreversible reactions entropy is lost to the surroundings as heat, as the entropy has increased. In any natural process, entropy tends to increase. The relation between Gibbs free energy (↑) and entropy is shown in the graph. The entropy of a perfect crystal of an element, or a compound, at 0 K is zero, by convention.

$$S = \int \frac{C_p}{T} \, dT,$$

so if C_p/T, (C_p = heat capacity at constant pressure), is plotted against T, then the entropy at a particular temperature is the area under the curve from 0 K to T. The graph ignores any physical transformation.

Gibbs-Helmholtz equation the equation states the relation between Gibbs free energy (p.112), enthalpy (p.112) and temperature, T, for a process at *constant pressure*:

$$\Delta G = \Delta H + T\left(\frac{\delta G}{\delta T}\right)$$

$$\frac{\delta G}{\delta T} = -\Delta S, \text{ so } \Delta G = \Delta H - T\Delta S.$$

A corresponding equation can be used with Helmholtz free energy for a process at constant volume:

$$\Delta A = \Delta U + T\ \frac{\delta A}{\delta T}$$

where A is Helmholtz free energy and U is internal energy.

gas constant the gas equation (p.148) uses the constant R in the equation $pV = nRT$. R is a constant for all gases and has the value $8.314\ \text{JK}^{-1}\ \text{mol}^{-1}$.

isothermal expansion describes the expansion of a given mass of gas at a constant temperature. If an ideal gas (p.147) expands against an external resistance, heat must be supplied to maintain a constant temperature. In practice, any change must take place very slowly to allow heat to flow either to, or from, the surroundings. The relation between pressure and volume is represented by $pV = k\ (= nRT)$.

adiabatic expansion[1] describes the expansion of a given mass of gas, in which no heat enters or leaves the system. An adiabatic compression leads to a rise in temperature and an adiabatic expansion to a fall in temperature. The relation between pressure and volume is represented by $pV^{\gamma} = \text{constant}$, where γ is the ratio C_p/C_v, i.e. the ratio of the heat capacities at constant pressure and constant volume. During an adiabatic change, the entropy of the system remains constant.

chemical potential the Gibbs free energy (p.112) per mole. It has the symbol μ. $\mu = \delta G/\delta n$, where n is the number of moles of a substance. For a chemical system, $G = \Sigma n_i \mu_i$, where μ_i is the chemical potential of a species, i, in a mixture with n_i moles of the species. This allows G to be calculated for a reaction. If $\Sigma n_i \mu_i = 0$, a closed system is in equilibrium.

Carnot's cycle the cycle has 4 stages: (1) isothermal expansion (↑) from A at temperature T_2 to B, heat energy enters system; (2) adiabatic expansion (↑) from

p/V for Carnot's cycle

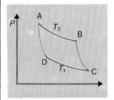

—— isothermal change
—— adiabatic change

Carnot's cycle

A	energy Q_2 in	B
	T_2 temperature constant	
$T_1 \rightarrow T_2$		$T_2 \rightarrow T_1$
entropy constant		entropy constant
D	energy Q_2 out	C
	T_1 temperature constant	

**graph of vapour pressure/
temperature for a liquid**
(diagrammatic)

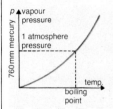

**graph of
log (vapour pressure)** $\Big/ 1/T$

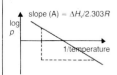

B to C, temperature falls to T_1; (3) isothermal compression from C to D, at temperature T_1, heat energy leaves the system; (4) adiabatic compression from D to A, temperature rises to T_2. If the energy absorbed is Q_2 and the energy given out is Q_1, then $w = Q_2 - Q_1$, where w is the work done by the system. The cycle is reversible, and using ideal gases, the efficiency of the cycle is:

$$\frac{w}{Q_2} = \frac{Q_2 - Q_1}{Q_2}.$$

The thermodynamic efficiency is usually given as: $(T_2 - T_1)/T_2$. This is the highest efficiency that can be obtained. Non-ideal gases produce a lower efficiency.

Clausius-Clapeyron equation the equation states the variation of the vapour pressure (p.154) of a pure liquid substance with the temperature. It is:

$$\frac{d}{dT}(ln\,p) = \frac{\Delta H_v}{RT^2}$$

where p is the vapour pressure, T the absolute temperature, R is the gas constant (↑) and ΔH_v is the molar latent heat (↓)of vaporization. It can also be written in the form:

$$\frac{dp}{dT} = \frac{p\Delta H_v}{RT^2}$$

If ΔH_v is assumed to be independent of temperature (true over small ranges of temperature) then, on integration,

$$\log p = A - \frac{\Delta H_v}{2.303RT}$$

This equation, with a suitable value for A, forms the curve of vapour pressure against temperature, *see diagram*. A graph of log p against $1/T$ is a straight line, and the slope is $-\Delta H_v/RT$.

latent heat the heat needed to change the state of matter of a substance at a constant temperature. The molar latent heat of fusion is the number of joules of heat required to change one mole of a solid to liquid, at a specified temperature, usually the melting point. The molar latent heat of vaporization is the number of joules of heat required to change one mole of a liquid to vapour, at a specified temperature, usually the boiling point of the liquid. Latent heat varies with temperature.

Trouton's rule an experimental rule which states that $\Delta H_v/T = 88$. This rule applies to non-polar liquids. Polar liquids, e.g. water, give a value of the constant higher than 88.

colligative properties those properties of a substance which depend upon the concentration of ions or molecules and are physical and not chemical in nature, e.g. changes in vapour pressure (p.154), osmotic pressure (p.119).

Raoult's law the relative lowering of the vapour pressure of a solvent, caused by the addition of a solute which does not dissociate into ions (p.124), is equal to the mole fraction of the solute in the solution. The law is true for dilute solutions, but does not hold for concentrated solutions; solutions that follow the law are called ideal solutions (↓), as most solutions show some deviation (↓). The equation for the law is:

$$\frac{\Delta p}{p^\circ} = \frac{n_B}{n_A + n_B}$$

where Δp is the lowering of the vapour pressure, p°, of the pure liquid by the addition of n_B moles of solute to n_A moles of solvent.

deviation (n) a failure to obey a law, rule, pattern, or principle. Many laws, rules, etc., are restricted, e.g. laws concerning solutions are true for dilute solutions, but as the concentration of solute rises, the law is no longer true and results show a deviation from the predicted result.

ideal solution a solution which obeys Raoult's law (↑) in all concentrations, preferably for the solution of two miscible (p.11) liquids. The total vapour pressure of such a solution is equal to the sum of the partial vapour pressures of both liquids. Such a solution shows no heat change on mixing the constituents. A better definition of an ideal solution is given by the change in the standard chemical potential (p.114) of a substance with its increasing concentration in a solution.

non-ideal solution a solution with colligative properties which deviate (↑) from those of an ideal solution (↑).

boiling-point composition the boiling-point composition curve for methanol and water is shown in *the diagram*. If a liquid mixture of composition A is boiled, at a temperature, T, under atmospheric pressure, the composition of the vapour is given by B, i.e. richer in methanol. Such a mixture has no maximum or minimum boiling point. Some mixtures have a maximum and some have a minimum boiling point, and these mixtures cannot be separated completely by distillation. All three types of mixtures deviate from an ideal solution (↑) as they deviate from Raoult's law (↑)

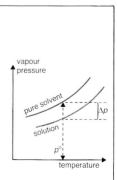

Raoult's law

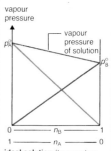

ideal solution (temperature constant)

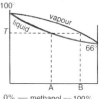

boiling point (°C) at 760 mm mercury

0% — methanol — 100%
100% — water — 0%

boiling point – composition curve

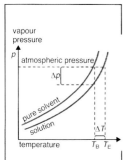

elevation of boiling point

elevation of boiling point the presence of a non-volatile solute in a pure liquid solvent raises the boiling point of the solvent. When the solute is added, the vapour pressure is lowered, so the solution boils at a higher temperature (T_B is the boiling point of the pure solvent and T_E is boiling point of the solution, both temperatures at atmospheric pressure). If ΔT is the elevation of the boiling point, θ_b is the boiling point constant and n is the mole fraction of the solute, then $\Delta T = \theta_b n$.

boiling-point constant this is a constant for a particular solvent; it has the symbol θ_b. Typical examples of θ_b are: water, 0.51°; benzene, 2.53°; propanone, 1.71°. If one mole of solute is dissolved in 1 kg of solvent, the predicted rise in the boiling point would be θ_b, i.e 0.51° for water. The formula for the elevation of the boiling point can be used for dilute solutions only, i.e. when the mole fraction (p.29) of solute is small compared with the mole fraction of the solvent. For dilute solutions

$$\theta_b = \frac{RT^2 M_A}{\Delta H_A \times 1000}$$

where R is the gas constant, T the boiling point of the pure solvent, M_A the molar mass of the solvent and ΔH_A the enthalpy of the solvent, at its boiling point.

ebullioscopic constant same as **boiling-point constant**

depression of freezing point the presence of a non-volatile solute in a pure solvent depresses, or lowers, the freezing point. The pure solvent freezes when the vapour pressure of the solid form is equal to the vapour pressure of the liquid form. When the solute is added, the vapour pressure of the solvent is lowered, and if pure solvent separates out, the freezing point is lowered. *see diagram* (T_f is the freezing point of the pure solvent, T_D is the freezing point of the solution). If ΔT is the depression of the freezing point, θ_f is the freezing point constant, and n is the mole fraction (p.29) of solute, then $\Delta T = -\theta_f n$.

depression of freezing point

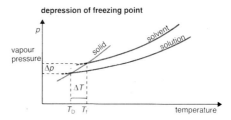

freezing-point constant this is a constant for a particular solvent; it has the symbol θ_f. Typical examples of θ_f are: water, 1.86°; benzene, 5.08°; naphthalene 6.9°; camphor, 37.7°. If one mole of solute is dissolved in one kg of solvent, the predicted depression of the freezing point is θ_f, i.e. 1.86° for water. The formula for the depression of the freezing point can be used for dilute solutions only, i.e. when the mole fraction (p.29) of solute must be small compared with the mole fraction of the solvent. For dilute solutions:

$$\theta_f = \frac{RT^2 M_A}{\Delta H_A \times 1000}$$

where R is the gas constant, T the freezing point of the pure solvent, M_A the molar mass of the solvent, and ΔH_A the enthalpy (p.112) of the solvent at its freezing point (i.e. the molar latent heat of freezing).

cryoscopic constant another name for **freezing-point constant** (↑).

eutectic point if two or more solids can dissolve in the liquid forms of each other, and each can lower the freezing point (or melting point) of the other, then they form a solid solution, and the lowest freezing point of any composition is the eutectic point. Such solids are not completely soluble in each other.

eutectic mixture the mixture of two solids which has the lowest freezing point of all possible compositions; it freezes at the eutectic point (↑). Eutectic mixtures are important in forming alloys with low melting points.

membrane (*n*) a very thin solid piece, or a porous piece, of a material which allows molecules and ions to pass through it.

permeable membrane a membrane (↑) which allows fluids to pass through it and also molecules and ions of solutes, but does not allow colloidal (p.174) particles or suspension particles to pass. *See* **dialysis** (p.177).

semipermeable membrane a membrane (↑) which allows molecules of some liquids to pass through, but not molecules and ions of solutes. Some varieties of cellophane and a membrane of copper (II) hexacyanoferrate (II) form semipermeable membranes.

osmosis (*n*) a process in which solvent molecules from a dilute solution pass through a semipermeable membrane (↑) to a concentrated solution. Osmosis tends to equalize the concentration of the solution separated by the membrane.

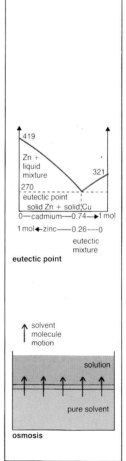

eutectic point

osmosis

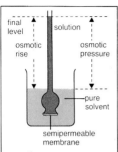

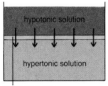

osmotic pressure

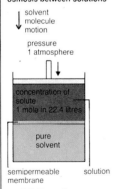

semipermeable
membrane

osmosis between solutions

solvent
molecule
motion

pressure
1 atmosphere

concentration of
solute
1 mole in 22.4 litres

pure
solvent

semipermeable   solution
membrane

van't Hoff equation

osmotic pressure the osmotic pressure of a solution is equal to the pressure required to prevent osmosis (↑) when the solution is separated by a semipermeable membrane (↑) from the pure solvent. The symbol for osmotic pressure is Π.

osmotic rise the rise in the level of a solution up a tube when osmosis (↑) drives the solvent through a semipermeable membrane (↑) into the solution. Osmosis stops when the pressure of the column of solution is equal to the osmotic pressure (↑).

isotonic (adj) describes two solutions which have the same osmotic pressure (↑), so osmosis does not take place when the solutions are separated by a semipermeable membrane.

hypotonic (adj) describes a solution which has a lower osmotic pressure than that of a given, or standard, solution. When the solutions are separated by a semipermeable membrane, solvent molecules will flow out of the hypotonic solution, e.g. a 0.05 M solution of glucose in water is hypotonic to a 0.1 M solution of glucose; water passes into the 0.1 M solution.

hypertonic (adj) describes a solution which has a higher osmotic pressure than that of a given, or standard, solution. When the solutions are separated by a semipermeable membrane, solvent molecules will flow into the hypertonic solution, e.g. 0.1 M glucose solution in water is hypertonic to a 0.05 M solution of glucose; water passes into the 0.1 M solution.

van't Hoff equation an equation which relates the osmotic pressure of a solution (Π) to the concentration in moles per metre^3 (c). The relation is: $\Pi = RTc$, where R is the gas constant, and T is the absolute temperature of the solution. Π is given in newtons per metre^2. If one mole of solute is dissolved in 22.4 litres of solvent, the osmotic pressure is 1 atmosphere (101.325 kNm^{-2}). This equation is valid only if the solute molecules neither dissociate (p.129), ionize (p.125), nor associate (p.118), and if the solution is dilute.

osmotic equation another name for **van't Hoff equation** (↑).

osmometer (n) an instrument for measuring osmotic pressure (↑). Modern instruments use membranes of cellulose, and are constructed from glass and steel to stand high osmotic pressures. Osmometers are used to determine the molar mass of polymers with macromolecules (i.e. very large molecules).

association (n) the joining together of two molecules to form a single molecule of twice the molecular mass. It is a reversible process as the associated molecule can dissociate (p.129), e.g. when in solution in benzene, ethanoic acid associates as follows:
$$2CH_3COOH \leftrightharpoons (CH_3COOH)_2.$$

degree of dissociation this refers to ionization in solution and to chemical dissociation of gases, e.g. $PCl_5 \rightarrow PCl_3 + Cl_2$. The symbol for the degree of dissociation is α, it is the fraction of one mole that dissociates. *See* **osmotic coefficient** ($\downarrow$).

van't Hoff factor usually designated by i.
$$i = \frac{\text{actual relative molecular mass}}{\text{observed value of relative molecular mass}}$$
The factor can be used to determine the degree of dissociation ($\uparrow$) or association ($\uparrow$).

osmotic coefficient let the degree of dissociation be α for the following dissociation:
$$AB \rightarrow A + B$$
before diss. 1 mol   0 mol 0 mol Total = 1 mol
after diss. $(1 - \alpha)$ mol α mol α mol Total = $(1 - \alpha)$ mol

$$\therefore \frac{\text{actual effect with dissociation}}{\text{expected effect if no dissociation}} = \frac{1 - \alpha}{1} = g$$

g is the osmotic coefficient. It is useful in the investigation of abnormal results for lowering of vapour pressure, elevation of the boiling point, depression of the freezing point, and osmotic pressures. The abnormal effects of colligative properties (p.116) are caused by dissociation, usually ionization, and the degree of dissociation can be measured in this way.

partition law when a solute is dissolved in two immiscible (p.11) solvents, A and B, the solute distributes itself between the two solvents according to the law:
$$\frac{\text{concentration in solvent A}}{\text{concentration in solvent B}} = K$$
where K is a constant, called the partition coefficient ($\downarrow$). The law is valid only for dilute solutions.

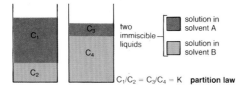

two immiscible liquids

solution in solvent A

solution in solvent B

$C_1/C_2 = C_3/C_4 = K$ **partition law**

partition coefficient the value of K, the partition coefficient, depends on temperature, and the actual solute used. The solute must not associate (↑) or dissociate (p.129) in either solvent, or else the partition law (↑) is not valid. If the solute associates in solvent B, so that n molecules form one associated molecule, then:

$$K = \frac{c_A}{n\sqrt{c_B}}$$

where c_A is the concentration in solvent A, and c_B the concentration in solvent B.

Henry's law the mass of a gas dissolved by a given volume of liquid at a constant temperature is proportional to the pressure of the gas. The law can also be stated as the volume of a gas dissolved by a given volume of liquid at constant temperature, and measured at the pressure used, is independent of that pressure. The law is not valid for concentrated solutions, neither is it valid if the gas associates or dissociates in the liquid, nor if the gas reacts chemically with the liquid, nor for high pressures.

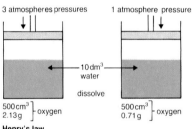

Henry's law

activity coefficient a factor by which the molecular concentration of a substance must be multiplied to make the concentration equal to the thermodynamic activity (↓) in a chemical system; its symbol is γ.

activity[2] (n) a thermodynamic quantity which is a measure of the effectiveness of a chemical substance in a particular system. If c is the molecular concentration, and a is the activity, then $a = vc$. In a non-ideal solution, the chemical substance behaves as though it had a concentration of vc instead of c.

active mass in the law of mass action (p.181), active mass means the concentration in moles multiplied by the activity coefficient (↑).

conductance (*n*) conductance = 1/resistance; it is measured in ohm⁻¹. A conductor with a resistance of 5 ohms has a conductance of 0.2 ohm⁻¹.

conductivity (*n*) the ability of a conductor to conduct electric current. Conductivity = 1/resistivity; its symbol is K, and its units are ohm⁻¹ metre⁻¹. The conductivity of a solution changes with concentration.

specific conductivity another name for **conductivity** (↑); it is no longer used.

molar conductivity the conductivity (↑) of a solution divided by the concentration in mol m⁻³. The symbol for molar conductivity is Λ and the units are ohm⁻¹ metre² mole⁻¹. Molar conductivity changes with dilution.

equivalent conductivity the conductivity (↑) of a solution multiplied by the volume (in m³) of solvent containing the mass of electrolyte which is equivalent to one faraday of electric charge, e.g. if the conductivity of a 0.01 M barium chloride solution is *x*, then the volume containing 0.5 mol of barium chloride is 50m³ (1 mol of barium chloride is equivalent to 2 faradays; 0.5 mol is equivalent to 1 faraday), so the equivalent conductivity is *x* × 50, while the molar conductivity is *x* × 100.

infinite dilution the dilution (p.127), or the concentration, at which molar conductivity (↑) reaches a maximum, as shown in the graph of molar conductivity against dilution. The molar conductivity at infinite dilution has the symbol Λ_∞. For strong electrolytes (p.124), it is the dilution, or concentration, at which molar conductivity becomes independent of concentration.

molar conductivity at infinite dilution *see* **infinite dilution** (↑). The graphs for molar conductivity against √concentration show that, for strong electrolytes, a value for Λ_∞ can be obtained by extrapolation, but no value can be obtained for weak electrolytes, as the graph is not a straight line.

Kohlrausch's law the equivalent conductivity (↑) at infinite dilution(↑) is equal to the sum of the ionic mobilities (↓) of the ions produced by the electrolyte. This law can also be stated as: the conductivity of an electrolyte is equal to the sum of the conductivities of the ions produced by the electrolyte when ionization is complete. As a formula: $\Lambda_\infty = \Lambda_{i+} + \Lambda_{i-}$, e.g. Λ_∞ KCl = Λ_i K⁺ + Λ_i Cl⁻, where Λ_∞ is the equivalent (or molar) conductivity, and Λ_i is the ionic mobility of an ion. The difference between molar and equivalent

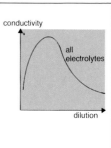

conductivity

all electrolytes

dilution

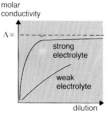

molar conductivity

Λ_∞

strong electrolyte

weak electrolyte

dilution

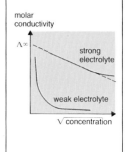

molar conductivity

Λ_∞

strong electrolyte

weak electrolyte

√ concentration

conductivities is shown by values for sodium sulphate:
molar conductivity:
$$\Lambda_\infty\, Na_2SO_4 = 2\Lambda_i\, Na^+ + \Lambda_i\, SO_4^{2-}$$
equivalent conductivity:
$$\Lambda_\infty\, Na_2SO_4 = \Lambda_i\, Na^+ + \tfrac{1}{2}\Lambda_\infty\, SO_4^{2-}.$$

ionic mobility the share of the molar (or equivalent) conductivity at infinite dilution carried by an ion; its symbol is Λ_i and it is a property of an ion. It is also a measure of the absolute velocity of an ion (p.124). Units are $ohm^{-1}\, m^2\, mol^{-1}$.

molar conductivity of an ion another name for **ionic mobility** (↑); it is preferable to use molar conductivity. Molar conductivities of ions are additive, e.g.
equivalent conductivity $\Lambda_\infty\, CaCl_2 = \tfrac{1}{2}\Lambda_i Ca^{2+} + \Lambda_i Cl^-$
molar conductivity $\quad \Lambda_\infty\, CaCl_2 = \Lambda_i Ca^{2+} + 2\Lambda_i Cl^-$
hence molar conductivity =
$\qquad\qquad$ 2× equivalent conductivity for $CaCl_2$.

transport number the fraction of the total current carried by a particular ion during electrolysis (p.126), i.e. the fraction carried by the anion (p.126) or the cation (p.126). The symbol for transport number is t. In the electrolysis of KCl, $t_{K+} + t_{Cl-} = 1$, that is, the transport number of the cation plus the transport number of the anion always equals unity. Transport number, molar conductivities at infinite dilution (↑) and molar conductivities of ions (↑) are related: $t_+\Lambda_\infty = \Lambda_{i+}$. For KCl, the data are:
$$\Lambda_\infty\, KCl = 1.3 \times 10^{-2}\, ohm^{-1}\, m^2\, mol^{-1}$$
$$K^+\!: \Lambda_i = 6.37 \times 10^{-3}\, ohm^{-1}\, m^2\, mol^{-1};\ t_{K+} = 0.49$$
$$Cl^-\!: \Lambda_i = 6.63 \times 10^{-3}\, ohm^{-1}\, mmol^{-1};\ t_{Cl-} = 0.51$$

now $0.49 \times 1.3 \times 10^{-2} = 6.37 \times 10^{-3}$
$\qquad 0.51 \times 1.3 \times 10^{-2} = 6.63 \times 10^{-3}$
and $0.49 + 0.51 = 1$.

velocity of ions the absolute velocity of an ion is defined as its velocity in metres per second when moving under a potential gradient of 1 volt per metre. The symbol for the velocity of an ion is υ. The equivalent conductivity at infinite dilution is given by:
$$\Lambda_\infty = F(\upsilon_+ + \upsilon_-) = 96\,500\,(\upsilon_+ + \upsilon_-)$$
Also $\dfrac{\text{speed of cation}}{\text{speed of anion}} = \dfrac{\upsilon_+}{\upsilon_-} = \dfrac{t_+}{t_-}$
where t is the transport number (↑). Hence for KCl,
$$\upsilon_{K+} = \frac{\Lambda_\infty \times {}^tK_+}{96\,500} = \frac{1.3 \times 10^{-2} \times 0.49}{95\,500}$$
$$= 6.6 \times 10^{-8}\, ms^{-1}$$

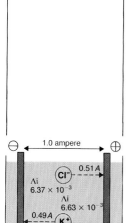

1.0 ampere

Cl^- $\quad$ 0.51 A
Λ_i 6.37×10^{-3}
Λ_i 6.63×10^{-3}
0.49 A $\quad K^+$
$\Lambda\infty = 1.30 \times 10^{-2}$

potassium chloride solution at infinite dilution

molar conductivities and transport numbers

transference number an alternative term for **transport number** (p.121).

conductivity cell a cell used to measure the conductivity (p.122) of a solution; two types of cell are illustrated. The cell contains two electrodes of stout platinum, generally coated with a thin layer of platinum black. The resistance of the volume of solution between two electrodes is measured by a conductance bridge. Solutions are prepared with conductance water (↓).

conductance cell see **conductivity cell** (↑).

cell constant the resistance of a solution in a conductivity cell (↑) is directly proportional to the distance apart of the electrodes, l, and inversely proportional to the area of the electrodes, a. as the electrodes of a cell are rigidly fixed, these dimensions never alter, and form a cell constant:

$$\text{conductivity} = \frac{1}{R} \times \frac{1}{a} = \frac{1}{R} \times \text{cell constant}$$

The cell constant is measured by using a solution whose conductivity is known, e.g. 0.1 M KCl.

conductance water ordinary distilled water has a conductivity of $5–10 \times 10^{-4}$ ohm^{-1} metre^{-1}, which affects the conductivity of a solution. Specially purified water, called conductance water, is obtained by further purification of distilled water; it has a conductivity of $0.05 – 0.5 \times 10^{-4}$ ohm^{-1} metre^{-1}.

conductivity water the same as **conductance water** (↑).

electrolyte (n) any compound which in solution or in a molten state, conducts an electric current and is decomposed by the current. The solvent is usually water, but other suitable solvents may be used. Acids, bases, and salts are generally electrolytes.

non-electrolyte (n) any compound which does not behave as an electrolyte (↑). Most organic compounds are non-electrolytes.

ionic theory the basic ideas of the theory are: (1) a solution of an electrolyte (↑), or a molten electrolyte, contains free ions (↓); (2) the conductivity of a solution of an electrolyte depends on the number and speed of the free ions in the solution; (3) the extent of ionization depends on the dilution of a weak electrolyte but is complete for a strong electrolyte.

ion (n) an atom, or a group of atoms, which has lost or gained electrons and thus acquired an electric charge. Atoms are neutral and lose or gain electrons; these

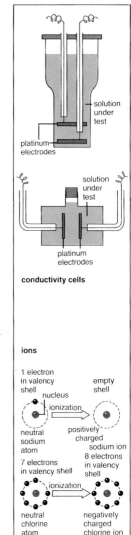

conductivity cells

ions

1 electron
in valency
shell   empty
     nucleus   shell
ionization
neutral   positively
sodium   charged
atom   sodium ion
           8 electrons
7 electrons   in valency
in valency shell shell
ionization
neutral   negatively
chlorine   charged
atom   chlorine ion

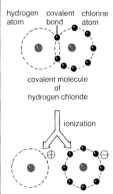

hydrogen covalent chlorine
atom   bond   atom

covalent molecule
of
hydrogen chloride

ionization

free ions in solution

positively   negatively
charged   charged
hydrogen   chlorine
ion   ion

chlorine atom attracts
one electron from a
hydrogen atom

ionization

electrons are taken from, or added to, the valency shell. Groups of atoms form the ions of acid radicals, or complex ions (p.144). Gaseous atoms may form ions. Solid electrolytes (↑) consist of ions combined in a lattice by ionic bonds (p.69). Loss of one or more electrons by an atom forms an ion with a positive charge; the gain of one or more electrons forms an ion with a negative charge. **ionic** (*adj*), **ionize** (*v*).

ionization (*n*) the formation of ions (↑) from atoms. Energy in the form of radiation, heat, or a high electrical potential produces ions from atoms. The atoms of some elements attract electrons (*see* **electronegativity**, p.75) to form ions; this action may cause ionization when a covalent compound is dissolved in water, e.g. hydrogen chloride in solution undergoes ionization as the chlorine atoms attract electrons from the hydrogen atoms. The dissolution of a solid electrolyte in a solvent separates the ions held in the crystal lattice (p.165), forming free ions; this process is not ionization.

ionic radii if the ions in a crystal are considered to be spheres, then the internuclear distances are the sum of the radii of the two ions. A table of empirical ionic radii can be constructed using data from the X-ray analysis of crystal lattices (p.165). Such a table was constructed by Pauling by assuming values for the radii of fluoride and oxide ions, since the X-ray data give only the internuclear distances. Electron density maps (p.69) offer further evidence. The table shows: (1) ions in a vertical group of the periodic table increase in size as atomic mass increases; (2) an ion of an element with a lower charge is larger than an ion of the element with a higher charge; (3) anions (p.124) are larger than the atoms from which they are formed; cations (p.124) are smaller than the atoms.

ionic radii

ion	radius $\times 10^{-10}$ m	ion	radius $\times 10^{-10}$ m
	0.60		0.84
	0.95		0.69
	1.33		0.56
	1.48		0.45

ionic and covalent radii

atom	covalent radius $\times 10^{-10}$ m	ion	ionic radius $\times 10^{-10}$ m
F	0.64	F⁻	1.36
Cl	0.99	Cl⁻	1.81
Br	1.14	Br⁻	1.95
I	1.33	I⁻	2.16

anion (*n*) an ion with a negative charge, which may be 1, 2 or 3 times as great as the charge on one electron.

cation (*n*) an ion with a positive charge, which may be 1, 2, 3, 4, 5, 6 or 7 times as great as the charge on one electron.

electrolysis (*n*) the decomposition of an electrolyte (p.124) in solution, or a molten electrolyte, by an electric current. Two electrodes, a positive anode and a negative cathode, are immersed in the electrolyte. Anions are attracted to, and discharged at, the anode; cations are attracted to, and discharged at, the cathode. The ions of the electrolyte complete the circuit between anode and cathode by supplying electrons to the anode and removing electrons from the cathode, in the process of discharging ions. A current of 1 amp corresponds to a flow of 625×10^{16} electrons per second. *See* **transport number** (p.123).

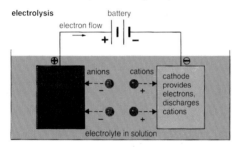

decomposition voltage the minimum voltage that must be applied to two electrodes in order to cause continuous electrolysis (↑) of an electrolyte (p.124) in solution. With some electrodes and electrolytes it is 0 volts. *See* **theoretical decomposition voltage** (p.198).

Faraday's laws these relate the products of electrolysis to the quantity of electric current used. Quantity of electric current is equal to the product of the current in amperes and the time in seconds; the product is the quantity measured in coulomb, e.g. 2 amps × 3 seconds = 6 coulomb. The laws state: (1) the amount of chemical change produced by an electric current is directly proportional to the quantity of electric current passed: (2) the amounts of different substances deposited or released by the same quantity of the electricity are proportional to their molar masses

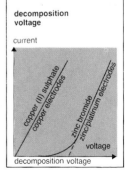

decomposition voltage

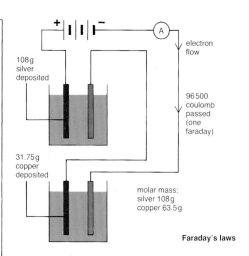

108 g silver deposited

electron flow

96 500 coulomb passed (one faraday)

31.75 g copper deposited

molar mass:
silver 108 g
copper 63.5 g

Faraday's laws

divided by the charge on their ions, e.g. 1.08 g silver and 0.3175 g copper are deposited by the same quantity of electricity: $Ag = 108$; Ag^+ discharged; $Cu = 63.5$; Cu^{2+} discharged; $Ag:Cu = 108/1 : 63.5/2$.

faraday (*n*) the quantity of electric current that forms or liberates one mole of monovalent ions, e.g. one faraday deposits 108 g silver or 31.75 g copper, as copper ions are divalent. One faraday of electric charge contains 1 mole of electrons. The approximate value of the faraday is 96 500 coulomb.

weak electrolyte an electrolyte (p.124) which in solution is a poor conductor owing to only a small fraction of its molecules being split into free ions. At great dilutions (↓) ionization is almost complete. Most organic acids and bases, ammonia solution and water are weak electrolytes.

strong electrolyte an electrolyte which in solution is a better conductor than a weak electrolyte (↑). This is because the electrolyte is almost entirely split into free ions in the solution. Free ions are also present in the molten state. In a solution, the number of ions present depends only on the concentration of the strong electrolyte.

dilution (*n*) the dilution of a solution is the reciprocal of its concentration in moles per litre; it is usually defined as the number of litres containing 1 mole of solute.

dilution in litres	conductivity ohm⁻¹ metre⁻¹	
	potassium chloride	ethanoic acid
0	0	0
1	9.82	0.132
10	1.12	0.046
100	0.12	0.014
1000	0.012	0.0041
10000	0.0013	0.0011

conductivity of weak and strong electrolytes

Ostwald's dilution law

concentration c (× 10^8) mol litre^-1	dilution V litre mol^-1	$\sqrt{V}$	α	$\dfrac{α}{\sqrt{V}}$ (×10^3)	K'_a (× 10^5)
0.02801	35700	188.94	0.5393	2.854	1.768
0.11135	8980.7	94.767	0.3277	3.458	1.779
0.21844	4577.9	67.66	0.2477	3.660	1.781
1.02831	972.47	31.184	0.1238	3.970	1.797
2.4140	414.25	20.353	0.0829	4.073	1.809
5.91153	169.16	13.006	0.0540	4.152	1.823
9.8421	101.60	10.079	0.0422	4.189	1.832
20.000	50.00	7.071	0.02987	4.224	1.840
52.303	19.119	4.3725	0.01865	4.265	1.854

α **proportional to** $\sqrt{V}$

$\therefore \quad \dfrac{α}{\sqrt{V}} \approx$ **constant**

Ostwald's dilution law for a weak binary electrolyte, the degree of ionization is proportional to the square root of the dilution. Let the degree of ionization be α and the dilution V. Let the ionization be AB $\rightleftharpoons$ A$^+$ + B$^-$, then by the law of mass action (p.181):

$$K'_a = \frac{[A^+][B^-]}{[AB]} = \frac{[α/V][α/V]}{[(1-α)/V]} = \frac{α^2}{(1-α)V}$$

If α is small then $(1-α)$ is approximately 1, and so:
$α^2 \simeq K'_a V$ or $α \quad \sqrt{K'_a V}$
i.e. α is proportional to $\sqrt{V}$. Ostwald's law is only approximately true, becoming more accurate for very dilute solutions of weak electrolytes; it is not applicable to strong electrolytes.

dissociation constant[1] in the expression for Ostwald's dilution law (↑), K'_a is the dissociation constant. K'_a alters with change in dilution.

ionization constant the equilibrium constant for the dissociation of a weak electrolyte can be defined in terms of activities (p.121). The constant is K_a.

$$K_a = \frac{aA_+ \times aB_-}{a_{AB}} = \frac{γ_+[A^+] \times γ_-[B^-]}{γ_{AB}[AB]} = \frac{K'_a \, γA_+ \times γB_-}{γ_{AB}}$$

where ν is the activity coefficient of a species and K'_a is the dissociation constant (↑). At infinite dilution, all activity coefficients become 1, so $K_a = K'_a$. A graph of log K'_a against $\sqrt{αc}$, where α is the degree of ionization and c is the concentration, is a straight line, hence the value of K_a can be extrapolated.

ion-pair[2] (n) electrostatic attraction between ions of opposite charge forms *ion-pairs*; these behave as if they were un-ionized molecules and do not carry an

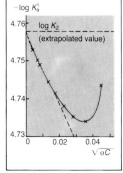

the dissociation and ionization constants

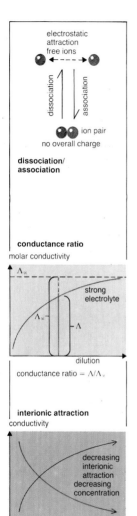

electrostatic
attraction
free ions

dissociation

association

ion pair
no overall charge

**dissociation/
association**

conductance ratio
molar conductivity

Λ_∞

strong
electrolyte

Λ_∞

Λ

dilution

conductance ratio = Λ / Λ_∞

interionic attraction
conductivity

decreasing
interionic
attraction
decreasing
concentration

dilution

**the factors affecting
conductivity in a
strong electrolyte**

electric current, e.g. a hydrogen ion and an ethanoate ion form an ion-pair which behaves similarly to a molecule of ethanoic acid.

dissociation (n) the splitting of an ion-pair (↑) into free ions capable of carrying an electric current. Dissociation thus refers to only those ions free to carry current. Compare **ionization** which applies to all ions whether free or in ion-pairs. An electrolyte can be completely ionized, but incompletely dissociated. Ordinary conductance experiments cannot distinguish between non-ionized molecules and ion-pairs.

association (n) the formation of ion-pairs (↑) from free ions, the opposite process to dissociation (↑).

ion conductance the molar conductivity of an ion at infinite dilution (p.122). It is given the symbol Λ_+ or Λ_- for cations and anions respectively. Ion conductances can be used to find the molar conductivities at infinite dilution of weak electrolytes. The value of Λ_∞ for ethanoic acid, HE t, is found from:

$$\Lambda_{H+} + \Lambda_{E-} = (349.8 + 40.9) \times 10^{-4}$$
$$= 390.7 \times 10^{-4} \text{ohm}^{-1}\text{m}^2$$

conductance ratio the ratio of the molar conductances of an electrolyte in solution at a given concentration to that at infinite dilution, i.e. the ratio Λ / Λ_∞. For weak electrolytes, the degree of dissociation, α, is given by $\alpha = \Lambda / \Lambda_\infty$; the conductance ratio does not give the degree of dissociation for strong or intermediate electrolytes.

interionic attraction as the concentration of an electrolyte in solution increases, the ions come closer together, and ions with opposite charges attract each other. This results in a decrease in the speed of the ions travelling towards an electrode and consequently a decrease in molar conductivity. This theory of interionic attraction offers an explanation of the conductance ratio for strong electrolytes, and the effect of dilution on the conductivity of a strong electrolyte.

ionic atmosphere the ions surrounding a particular ion in a solution. The ionic atmosphere has a resultant charge opposite in sign to the particular ion.

electrophoretic effect under an applied voltage, an ionic atmosphere (↑) moves in the opposite direction to that of a particular ion. The ionic atmosphere tends to drag the ion with it. This is equivalent to an increase in viscosity of the solution which slows down the speed of an ion towards an electrode.

degree of dissociation that fraction of an electrolyte which is dissociated into ions that are free to carry electric current; its symbol is α. In weak electrolytes, undissociated ions reduce α; in strong electrolytes, interionic attraction reduces α.

degree of ionization under Arrhenius' theory of ionization, no distinction was made between un-ionized molecules and ion-pairs; experimental results also make no distinction. The degree of dissociation was thus taken to be the degree of ionization and the degree of ionization calculated from Λ/Λ_∞, the conductance ratio.

Kohlrausch's equation an empirical equation to fit the conductivity data for strong electrolytes. It is: $(\Lambda_\infty - \Lambda)\sqrt{V}$ = constant, where Λ is the molar conductivity and V the dilution of a strong electrolyte. The relation is reasonably accurate at high dilutions.

ionization of water even the purest water acts as a weak conductor, so it is ionized to a small extent. The process is: $H_2O \rightleftharpoons H^+ + OH^-$. From the law of mass action:

$$K = \frac{[H^+][OH^-]}{[H_2O]}$$

The concentration of water molecules, H_2O, is very large, and can be taken as constant, so $K_W = [H^+][OH^-]$. See **ionic product** ($\downarrow$).

ionic product the value of $[H^+][OH^-]$ for the ionization of water ($\uparrow$) is called the ionic product, symbol K_W. Its value varies with temperature; at 25°C, $K_W = 1 \times 10^{-14}$. As water is neutral, $[H^+] = [OH^-]$. Concentrations, and not activities (p.121), can be used since the degree of dissociation ($\uparrow$) is very small. Hence, in water, $[H^+] = [OH^-] = \sqrt{K_W} = 1 \times 10^{-7}$ at 25°C.

K_w the ionic product of water. For most purposes $K_w = 10^{-14}$

hydration of ions the transport number (p.123) of a selection of cations in electrolytes with the same anion, all at 25°C, show that the smallest ions do not travel

ionic product of water

temperature	$K_w \times 10^{14}$
0°C	0.113
10°C	0.292
20°C	0.681
25°C	1.008
30°C	1.468
40°C	2.917
50°C	5.474

cation	transport number at 25°C	
	0.01 M	0.02 M
Li^+ in LiCl	0.329	0.326
Na^+ in NaCl	0.392	0.390
K^+ in KCl	0.490	0.490
½ Ba^{2+} in $BaCl_2$	0.440	0.437

speed $Li^+ < Na^+ < K^+$

effect of hydration on ions

hydration of ions

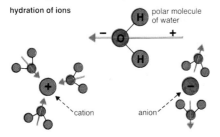

polar molecule of water

cation

anion

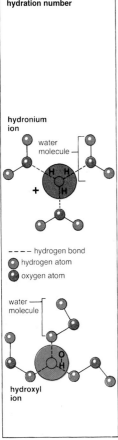

Li$^+$	5±1	F$^-$	4±1
Na$^+$	5±1	Cl$^-$	1±1
K$^+$	4±1	Br$^-$	1±1
Rb$^+$	3±1	I$^-$	1±1
Mg^{2+}	15±2		
Al^{3+}	26±5		

hydration number

hydronium ion

water molecule

+

H   H
H

- - - - hydrogen bond
⬤ hydrogen atom
⬤ oxygen atom

water molecule

O
H

hydroxyl ion

faster than the largest ions, as would be expected. This result is explained by the hydration of ions. Water molecules have a polar nature (p.75). Electrostatic attraction exists between an ion and a polar molecule. Solute ions are surrounded by oriented water molecules in one or more layers. This forms a larger particle than the naked ion. The attractive force increases with increasing electrovalency, and decreases with increasing radius, of an ion. Lithium ions thus have a greater attraction than sodium ions, and potassium ions, for water molecules. A hydrated lithium ion is larger than a hydrated sodium ion, and travels more slowly. Hydrated ions are shown as: Na$^+_{(aq)}$; Ba$^{2+}_{(aq)}$.

aquation (n) some water molecules form bonds with an ion, producing complex ions (p.144), e.g. [Cu(H$_2$O)$_4$]$^{2+}$. These ions, in turn, may be surrounded by oriented water molecules of hydration (↑). There is a gradation in bond strength between ion and water molecules, from coordinate bonds (p.77) to weak electrostatic attraction, as distance from the ion increases. All such forms of combining, or of associating, water molecules with an ion is aquation.

solvation (n) any solvent with polar molecules can form combinations, or associations, of solvent molecules with ions. The process is solvation. Aquation (↑) is solvation with water molecules.

aqua-ion (n) a complex ion with water molecules bonded to a cation by coordinate bonds, e.g. (Fe(H$_2$O)$_6$)$^{3+}$; (Al(H$_2$O)$_6$)$^{3+}$.

hydration number the number of water molecules associated with an ion in solution. The hydration number for a particular ion is variable, and depends on the experimental method used to evaluate the number. See **hydration of ions** (↑) for a description of the attractive force of hydration.

hydronium ion a hydrogen ion, or proton, is too small to exist by itself in the presence of water; it combines with a water molecule to form a hydronium ion, H$_3$O$^+$. The ionization of water is written more accurately as: H$_2$O + H$_2$O ⇌ H$_3$O$^+$ + OH$^-$. The hydronium ion is also hydrated, the most common ion being H$_9$O$_4^+$. The ion is bonded with water molecules using hydrogen bonds.

hydroxonium ion same as **hydronium ion** (↑).

hydroxyl ion the hydroxyl ion, OH$^-$, is hydrated, the most common ion being H$_7$O$_4^-$. The hydrogen atoms of the coordinating water molecules form hydrogen bonds.

enthalpy of hydration for sodium chloride the enthalpy of hydration is the heat of reaction of:

$$Na^+(g) + Cl^-(g) \rightarrow Na^+(aq) + Cl^-(aq)$$

with the ions in solution at infinite dilution. The ΔH value for the reactants is calculated from $(S + D/2 + I - E)$, where S = heat of sublimation, I = heat of ionization, of sodium, $D/2$ = heat of dissociation and E = electron affinity, of chlorine. ΔH_f is the heat of formation of crystalline sodium chloride, and ΔH_{soln} is the heat of solution at infinite dilution. *See* **Born-Haber cycle** (p.173). From these values, the heat of crystallization, i.e. the lattice energy, ΔH_{cryst}, and ΔH_{hyd}, the enthalpy of hydration, are calculated. ΔH_{hyd} is only slightly less than ΔH_{cryst}, and it supplies the energy to disrupt the crystal lattice and dissolve the solute. All enthalpies of hydration differ from the lattice energies of crystalline compounds by a small amount. If $\Delta H_{cryst} > \Delta H_{hyd}$, then absorption of heat occurs on forming a solution, and solubility increases with temperature rise. For anhydrous salts, there is a possibility that $\Delta H_{hyd} > \Delta H_{cryst}$ so heat is evolved on forming a solution, and solubility decreases with temperature.

Arrhenius' theory the ionic theory of the dissociation of electrolytes in solution. By this theory, the properties of acids are due to the presence of hydrogen ions. Modern theories are those of Brønsted-Lowry and Lewis.

hydrogen ion concentration for strong acids, the hydrogen ion concentration is equal to the concentration of the acid, e.g. in 2 M HCl, $[H^+] = 2$ M; in 0.1 M H_2SO_4, $[H^+] = 0.2$ M. For weak acids, the hydrogen ion concentration is less than the concentration of the acid, owing to the acid being only partially dissociated. Hydrogen ion concentration is also used with bases. In all aqueous solutions, the dissociation constant for water, K_w, remains constant and is taken to be 10^{-14}. For strong bases, the concentration of hydroxyl ions is equal to the concentration of the base. For 0.1 M NaOH, $[OH^-] = 10^{-1}$, hence $[H^+] = 10^{-14} \div 10^{-1} = 10^{-13}$. For weak bases, the concentration of hydroxyl ions is less than that of the base.

pH value for an acid or base solution, $pH = -\log_{10}[H^+] = \log_{10} 1/[H^+]$, where $[H^+]$ is expressed in moles per litre. If $[H^+] = 0.2$, then $1/[H^+] = 5$ and $pH = \log 5 = 0.699$; if $[H^+] = 1$, $pH = \log 1 = 0$.

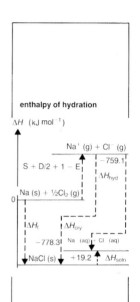

enthalpy of hydration

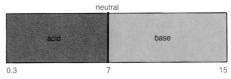

normal range of pH values

For a base, pH $= -\log_{10} K_w/[OH^-]$. For 0.1 M NaOH, $[OH^-] = 0.1$, pH $= -\log 10^{-14}/10^{-1} = -\log 10^{-13} = 13$; if $[OH^-] = 1$, pH $= 14$. For a neutral solution $[H^+] = [OH^-] = 10^{-7}$, so pH $= 7$.

pH meter an instrument which measures pH values (↑) directly. It consists of a glass electrode (↓) and a calomel electrode (↓) which dip into a solution of unknown hydrogen ion concentration.

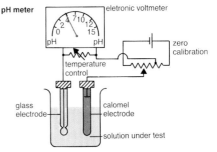

pH meter

glass electrode an electrode which is reversible with respect to hydrogen ions. A thin-walled glass bulb contains an approximately 0.1 M hydrochloric acid solution. A silver-silver chloride electrode dips into the acid. When placed in a solution of unknown pH, a potential difference is formed across the glass membrane, dependent on the concentration or, more accurately, the activity, of the hydrogen ions in solution. The resistance of the glass membrane is high, and the electrode can be used only with an electronic voltmeter.

calomel electrode the structure of this electrode is shown in *the diagram*. It is used as a secondary **standard electrode**, the hydrogen electrode (p.192) being the primary standard. The potential of the electrode remains constant at constant temperature. The electrode shown is a saturated calomel electrode and at 25°C its potential is +0.242V, on the IUPAC convention.

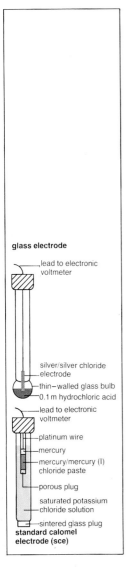

glass electrode

— lead to electronic voltmeter

— silver/silver chloride electrode
— thin-walled glass bulb
— 0.1 m hydrochloric acid

— lead to electronic voltmeter

— platinum wire
— mercury
— mercury/mercury (I) chloride paste
— porous plug
— saturated potassium chloride solution
— sintered glass plug

standard calomel electrode (sce)

ion-selective electrode the construction of an ion-selective electrode is shown in the diagram. It has a plug of material which allows a limited number of ionic species to pass through; in the electrode illustrated, the plug is made of lanthanum fluoride and conducts only fluoride ions. Such an electrode replaces a glass electrode in a meter, and measures the concentration of fluoride ions.

amphoteric electrolyte an electrolyte which contains separate acidic and basic groups in the same molecule, e.g. an amino acid. The acidic ionization of an amino acid is represented by:

$^+NH_3.CH_2.COO^- + H_2O \rightleftharpoons H_3O^+ + NH_2CH_2COO^-$

the basic ionization by:

$^+NH_3CH_2COO^- + H_2O \rightleftharpoons {}^+NH_3CH_2COOH + OH^-$.

The ionic component of a solution is determined by the addition of an alkali or acid.

dissociation constant[2] weak acids and weak bases are both weak electrolytes. Ostwald's dilution law (p.128) can be applied to determine the degree of dissociation, and hence the hydrogen ion or hydroxyl ion concentration respectively. As the extent of dissociation is small, the ionization constant is used and not the dissociation constant (p.128). As an approximation $K_a = \alpha^2/V$. For a weak acid, the ionization constant is K_A, and $[H^+] = [A^-]$ for the reaction $HA \rightleftharpoons H^+ + A^-$. Now $\alpha \approx \sqrt{K_AV}$, and $[H^+] = \alpha \times$ concentration $= \alpha/V$, so

$$[H^+] \approx \frac{\sqrt{K_AV}}{V} = \sqrt{K_A/V}$$

A more accurate expression is $[H^+] = \frac{1}{2}K_A + \sqrt{K_A/V}$, used at very high dilutions. Similarly for a base in solution:

$K_B = \alpha^2/V$ and $[OH^-] = \alpha/V$, so $[OH^-] = \sqrt{K_B/V}$. K_W, the ionization constant for water (p.128), is related to $[OH^-]$ by $[H^+] = K_W/[OH^-]$. So in a solution of a base, $[H^+] = K_W/\sqrt{K1_B/V}$. For a weak acid, e.g. HCN, the reaction is:

$CN^- + H_2O \rightarrow HCN + OH^-, K_B = \dfrac{[HCN][OH^-]}{[CN^-][H_2O]}$

hence $\dfrac{K_W}{K_B} = \dfrac{[H^+][OH^-]}{[H_2O]} \times \dfrac{[CN^-][H_2O]}{[HCN][OH^-]} = \dfrac{[H^+][CN^-]}{[HCN]}$

$= K_A$

and $K_W = K_A \times K_B$. So dissociation constants for acids and bases can be put on the same scale, all given a value for K_A.

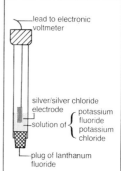

an ion-selecting electrode
(selecting for fluoride ions)

- lead to electronic voltmeter

silver/silver chloride electrode

solution of { potassium fluoride / potassium chloride }

plug of lanthanum fluoride

calculation of H⁺ concentration and pH

0.0001 M HCN

acid/base	pK	K_A	K_B
NH_4OH	9.24	5.75×10^{-10}	1.74×10^{-5}
CH_3NH_3OH	10.62	2.40×10^{-11}	4.17×10^{-4}
$C_2H_5NH_3OH$	10.63	2.34×10^{-11}	4.27×10^{-4}

calculation of OH⁻ and H⁺ concentration

0.1 M ammonia solution	
$k_B = 1.8 \times 10^{-5}$	$K_A = K_W/K_B$
$V = 10$ litres	$= 10^{-14} \div 1.8 \times 10^{-5}$
$[OH^-] = \sqrt{1.8 \times 10^{-5} \times 10^{-1}}$	$= 5.56 \times 10^{-10}$
$= 1.342 \times 10^{-3}$	$[H^+] = \sqrt{K_W K_A V}$
$[OH^+] = \dfrac{K_W}{[OH^-]} = \dfrac{10^{-14}}{1.342 \times 10^{-3}}$	$= \sqrt{10^{-14} \times 5.56 \times 10^{-10} \times 10}$
$= 7.46 \times 10^{-12}$	$= 7.46 \times 10^{-12}$
$pK_B = -\log k_B = 4.75$	$pH = -\log [H^+]$
$\therefore pK_A = 9.25$	
$pH = \frac{1}{2} K_W + \frac{1}{2} pK_A - \frac{1}{2} \log V$	$= -\log (7.46 \times 10^{-12})$
$= 7 + 4.625 - 0.5$	$= 11.13$
$= 11.13$	

ionization exponent ionization constants of acids and bases and the ionic product of water (p.130) can be expressed in a manner similar to pH values:
$pK_A = -\log K_A$; $pK_W = -\log K_W = 14$;
$pK_B = -\log K_B$. Since $K_W = K_A \times K_B$, hence
$pK_A + pK_B = pK_W = 14$. See **dissociation constant** (↑).
Now $-\log (H^+) = -\log \sqrt{K_A/V} = -\frac{1}{2} \log K_A + \frac{1}{2} \log V$,
so $pH = \frac{1}{2} pK_A + \frac{1}{2} \log V$ for acid solutions. For solutions of bases:
$$[H^+] = K_W \sqrt{\frac{V}{K_B}} = K_W \frac{\sqrt{VK_A}}{K_W} = \sqrt{K_W K_A} V$$
$-\log (H^+) = -\frac{1}{2} \log K_W - \frac{1}{2} \log K_A - \frac{1}{2} \log V$,
i.e. $pH = 7 + \frac{1}{2} pK_A - \frac{1}{2} \log V$.
pK value the ionization exponent (↑) of an acid or base.

indicator (*n*) a substance which varies its colour according to the hydrogen ion concentration of the solution in which it is placed. Acid/base indicators are generally weak organic acids. If the acid is HIn, the indicator reaction is: $HIn + H_2O \rightleftharpoons H_3O^+ + In^-$. The two species, HIn and In^-, have different coloured forms, and the colour exhibited depends on the hydrogen ion concentration of the solution. The equilibrium position between the two coloured forms is governed solely by the indicator exponent (↓) and pH values of the solution.

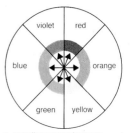

complimentary colours are opposite each other

coloured filters

indicators

| indicator | colour | | pK_{In} | pH range |
	acid	alkali		
methyl orange			3.7	3.1 – 4.4
methyl red			5.1	4.2 – 6.3
bromothymol blue			7.0	6.0 – 7.6
phenolphtalein	colour-less		9.4	8.3 – 10.0

indicator exponent the indicator constant for the reaction of an indicator (↑) is:

$$K_{In} = \frac{[H^+][In^-]}{[HIn]}$$

and $pK_{In} = -\log K_{In}$, where pK_{In} is the indicator exponent. $[H^+] = K_{In} \times [HIn]/[In^-]$ so $pH = pK_{In} + \log [In^-]/[HIn]$. A colour change is considered observable when either of the two species has a relative amount of 90%. The minimum pH producing a colour change is given by: $pH = pK_{In} + \log 10/90 = pK_{In} - 1$. The corresponding maximum change is $pH = pK_{In} + 1$. This gives the useful range of an indicator. (*See table*).

colorimeter (*n*) an instrument which discriminates a particular colour from other colours, used in connection with indicators (↑). **colorimetric (adv), colorimetry (n).**

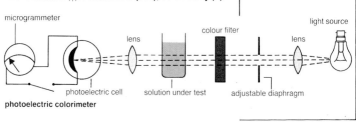

microgrammeter

light source

colour filter

lens

lens

photoelectric cell solution under test adjustable diaphragm

photoelectric colorimeter

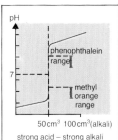

50 cm³ 100 cm³(alkali)

strong acid – strong alkali

50 cm³ 100 cm³(alkali)

weak acid – strong alkali

titration curves for 0.1 m alkali with 0.1 m acid

equivalence point and end point addition of strong alkali to weak acid

titrimetry (*n*) the measurement of the concentration of a solution, or an ionic species, by the process of running one solution into another solution until an equivalence point (↓) is reached. The result determines the volume of each solution at the equivalence point, and the required concentration is determined from the measured volume. **titrimetric** (*adj*).

titrand (*n*) the solution to be titrated. A measured volume (by pipette) has an indicator (↑) added to it.

titrant (*n*) the solution added to the titrand (↑).

titration (*n*) the reaction using measured amounts of titrand (↑) and titrant (↑). The process produces a titration value (↓).

titration value the volume of titrant added to reach the equivalence point (↓).

titre (*n*) a concentration of the titrant (↑) solution; the volume of titrant added.

titration curve (1) a graph which shows the change in pH with the addition by volume of a titrant (↑) in acid/alkali titrimetry. (2) a similar graph indicating the course of a titration in complexometric, precipitation or redox titrations. For acid/alkali titrations, different forms of the graph are obtained for strong or weak electrolytes. For strong acids titrated with strong alkalis there is a sudden change in pH from 4 to 10 when neutralization is reached. With weak electrolytes, the change in pH is less, and is not around a pH of 7.

equivalence point the point at which the stoichiometric amounts of titrand and titrant are equal, e.g. 50 cm³ of 0.1 M acid is equivalent to 50 cm³ 0.1 M alkali, as the amounts of acid and alkali in these solutions are proportional to the molar amounts of the substances in the equation for the reaction.

end-point (n.) the point in the titration at which an indicator shows a change from one set of conditions to another set of conditions, e.g. from an acidic to an alkaline solution. The end-point is labelled on the graph. A bad choice of an indicator will give an end-point which is not the same as the equivalence point, e.g. if methyl orange is used with weak acid/strong alkali titrations.

neutral point the point in a titration at which the pH has a value of 7. From the titration curves, it can be seen that the neutral point and the equivalence point (↑) do not always coincide. The natural point, equivalence point and end-point (↑) coincide for titrations of strong acid/strong alkali.

relative precision the relative precision of an end-point
is defined as the steepness of the titration curve around
the end-point. For acid-base titrations the rp is the
fraction of the stoichiometric titrant required to produce
a change of ± 0.1 pH on either side of the end-point.

potentiometric titration the titrand (p.135) is placed in a
vessel and two electrodes dipped into the solution. One
is a compact standard calomel electrode (p.133) and
the other is an indicator electrode; the latter is usually
an ion-selective electrode (p.134) suitable for the ion to
be titrated. The titrant is added in suitable increments,
and the e.m.f. of the cell measured by a potentiometer.
A graph is plotted of e.m.f. against the volume of titrant
added. The end-point is at the point of maximum slope.
A more accurate determination of the end-point is
achieved by plotting the ratio of $\Delta E / \Delta V$, the change in
e.m.f. for the addition of a definite small volume of
titrant. If ΔV is small, $\Delta E / \Delta V$ is a close approximation of
dE/dV, and the highest point on the graph gives the
end-point. Potentiometric titrations are very useful for
the titrations of coloured solutions, and very dilute
solutions.

electrometric titration another name for **potentiometric
titration** ($\uparrow$).

conductimetric titration the change in conductance
(p.122) can be used to follow the course of a titration,
particularly for acid-base reactions. As the hydrogen
ion or hydroxyl concentration falls with the addition of a
titrant, so the conductance decreases. It becomes a
minimum at the equivalence point (p.135), and then
rises again as more titrant is added. The method can be
used for the titration of weak acids against weak bases.
A conductance bridge is used to take measurements.

precipitation titration an analytical procedure in which
the chemical process of the titration (p.137) leads to
one of the reaction products being precipitated from
solution, e.g. the titration of silver nitrate solution with
sodium chloride solution leads to the precipitation of
insoluble silver chloride.

$$AgNO_3 + NaCl \longrightarrow AgCl + NaNO_3$$

thermometric titration an analytical procedure in which
the progress of the titration (p.137) and the determination
of the end-point (p.137) are carried out by measuring

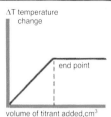

thermometric
titration

acid base indicator

phenolphthalein – an acid
base indicator with a
colour change at pH9

adsorption indicator

fluorescein – an adsorption
indicator

the temperature of the solution as the titrant is added.
The process has to be carried out in a thermally
insulated vessel and a graphical plot of the temperature
change against the volume of titrant added is made.

photometric titration a process in which changes in the
absorption of visible or ultraviolet radiation is used as
the basis for determining the progress and end-point
(p.137) of a titration (p.137), e.g. copper (II) ions can be
titrated with EDTA (p.208) at 745 nm as the copper-
EDTA complex absorbs more strongly at this
wavelength than do the copper (II) ions.

acid-base indicator a chemical compound which will
change its structure and colour at a specific pH value
(p.132) in solution and can be used in small quantities
to indicate the completion of a reaction in an acid-base
titration (p.137).

complexation indicator a chemical compound capable
of reacting with metal ions to produce a coloured
product which gives a colour change when the metal
ions are reacted with a complexone (p.208) such as
EDTA (p.208) in a complexometric titration (p.137).

solochrome black
an indicator for
complexation reactions
complexation indicator

metal ion indicator *see* **complexation indicator** (↑).
redox indicator a chemical compound which produces
a colour change at the end-point (p.137) of a redox
titration due to a change in its own chemical structure.
Iron (II) complexes of 1, 10-phenanthroline derivatives
are amongst the best redox indicators.

redox indicator

ferroin – an indicator
for redox titrations
blue   red

precipitation indicator *see* **adsorption indicator** (↓)
adsorption indicator describes a chemical compound
which is adsorbed on the surface of the precipitate in a
precipitation titration (p.137) and changes colour at the
end-point (p.137) of the titration, e.g. eosin and
fluorescein are two indicators used for this purpose.

buffer solution a solution which resists a change in hydrogen ion concentration when either an acid or an alkali is added. A buffer solution consists of a weak acid and its salt, or a weak base and its salt. For a weak acid, HA, the addition of hydrogen ions from an acid results in: $H_3O^+ + A^- = H_2O + HA$, and the addition of hydroxyl ions from an alkali results in: $OH^- + HA = H_2O + A^-$. For a weak base, BOH, the reactions are: $B^+ + OH^- = BOH$, and $BOH + H_3O^+ = B^+ + H_2O$. The pH of a buffer solution is obtained from the Henderson-Hasselbalch equation ($\downarrow$).

---

pH of a buffer solution

solution: 0.1 M ethanoic acid
        0.1 M sodium ethanoate
[acid] = 0.1 [salt] = 0.1   pK = 4.75
$pH = pK + \log \frac{[salt]}{[acid]} = 4.75 + \log \frac{0.1}{0.1}$

        $= 4.75 + \log 1 = 4.75$

---

addition of alkali to buffer solution

buffer solution as above
pH = 4.75 $[H^+] = 1.78 \times 10^{-5}$
10 cm^3 M NaOH added to 1 litre solution
∴ 0.01 M OH$^-$ ions added
∴ 0.01 M H$^+$ ions removed
so 0.01 M ethanoate ions formed
   0.01 M ethanoic acid removed
$pH = pK + \log \frac{[salt]}{[acid]} = 4.75 + \log \frac{(0.1 + 0.01)}{(0.1 - 0.01)}$

    $= 4.75 + \log \frac{0.11}{0.09} = 4.75 + 0.09$

$= 4.84$

---

buffer solution with a required pH

required pH = 4.0
ethanoic acid and sodium ethanoate
solution used pK = 4.75
$pH = pK + \log \frac{[salt]}{[acid]}$

$4.0 = 4.75 + \log \frac{[salt]}{[acid]}$

$\log \frac{[salt]}{[acid]} = -0.75 \quad \frac{[salt]}{[acid]} = 0.178$

if acid is 0.2 M [salt] = 0.2 × 0.178 = 0.036 M

Henderson-Hasselbalch equation the ionization constant, K_A, for a weak acid, HA, is given by:

$$K_A = \frac{[H^+]\,[A^-]}{[HA]}$$

In a mixture of a weak acid and a metal salt of the acid, the salt will be completely ionized, and the acid weakly ionized. As an approximation, the acid radical ions will be derived solely from the salt. Rearranging the equation:

$$[H^+] = K_A \frac{[HA]}{[A^-]} = K_A \frac{[acid]}{[salt]}$$

Taking logs, pH = pK + log [salt] / [acid]. This is the Henderson-Hasselbalch equation. For a weak alkali and its salt:

$$K_B = \frac{[B^-][OH^-]}{[BOH]} \text{ so } [OH^-] = K_B \frac{[alkali]}{[salt]} = \frac{K_W}{[H^+]}$$

$$\text{Therefore } [H^+] = \frac{K_W}{K_B} \frac{[salt]}{[alkali]} = K_A \frac{[salt]}{[alkali]} \text{ and}$$

$$pH = pK + \log \frac{[alkali]}{[salt]}$$

For buffer solutions used in practice, the ratio of [SALT]/[ACID] or [ALKALI]/[SALT] lies between 10 and 0.1. This produces a pH between pK + 1 and pK − 1.

hydrolysis of salts a partial reversal of neutralization, or the prevention of complete neutralization, between an acid and a base when a salt is dissolved in water, due to the action of the solvent, water. Hydrolysis occurs when the acid or the base, or both, are weak electrolytes. The resulting solution is acidic for weak base/strong acid, and alkaline for weak acid/strong base.

general equation for the hydrolysis of a salt BA

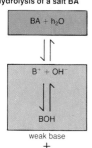

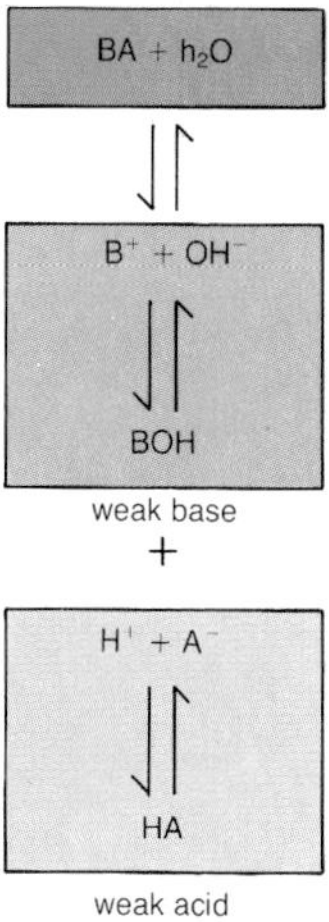

weak base
+

weak acid

solution of 0.1 M salt weak acid/strong alkali

$K_A = 1.8 \times 10^{-5}$ $\log K_A = -4.745$ $V = 10$

$pK = 4.745$ $\log V = 1$

$K_h = \dfrac{K_W}{K_A} = \dfrac{10^{-14}}{1.8 \times 10^{-5}} = 5.6 \times 10^{-10}$

$pH = -\frac{1}{2} \log K_W - \frac{1}{2} \log K_A - \frac{1}{2} \log V$

$= \frac{1}{2}(14 + 4.745 - 1) = 8.9$

$pH = 7 + \frac{1}{2} pK - \frac{1}{2} \log V$

$= 7 + 2.373 - 0.5 = 8.9$

lyolysis (n) the action of a polar solvent capable of acting as an acid or a base, or both, in partially reversing, or preventing, complete neutralization. Hydrolysis (p.141) is one form of lyolysis.

solvolysis (n) an alternative name for **lyolysis** (↑).

hydrolysis constant let the reaction between a salt and water be as shown. Let K_h be the hydrolysis constant. For a weak acid/strong base (*anion hydrolysis*):

$$B^+ + A^- + H_2O \rightleftharpoons HA + B^+ + OH^-$$

$K_h = \dfrac{[HA][OH^-]}{(A^-)}$ but $K_W = [H^+][OH^-]$ (*See* K_W, p.130).

therefore $K_h = \dfrac{K_W[HA]}{[H^+][A^-]} = \dfrac{K_W}{K_A}$ where K_A is the acid dissociation constant (p.128). For a weak base/strong acid (*cation hydrolvsis*):

$$B^+ + A^- + H_2O \rightleftharpoons BOH + A^- + H^+$$

Therefore $K_h = \dfrac{[BOH][H^+]}{[B^+]} = \dfrac{K_W[BOH]}{[B^+][OH^-]} = \dfrac{K_W}{K_B}$

where K_B is the base dissociation constant (p.128). Some weak bases form complex ions with water, so the calculation of K_h is more complicated than indicated here. For a weak base/weak acid:

$$B^+ + A^- + H_2O \rightleftharpoons BOH + HA,$$

therefore $K_h = \dfrac{[BOH][HA]}{[B^+][A^-]} = \dfrac{K_W[BOH][HA]}{[B^+][OH^-][H^+][A^-]}$

$= \dfrac{K_W}{(K_A \times K_B)}$.

degree of hydrolysis for a salt of a weak acid/strong base, the degree of hydrolysis, h, is given by:

$$A^- + H_2O \rightleftharpoons HA + OH^-$$
$$(1-h)/V \qquad h/V \quad h/V$$

so $K_h = \dfrac{[h/V] \times [h/V]}{[1-h]/V} = \dfrac{h^2}{[1-h]V}$

where K_h = hydrolysis constant. Since h is small, $(1-h) \approx 1$, hence $K_h = h^2/V$ or $h = \sqrt{K_h V}$, now $[OH^-] = h/V$

therefore $[H^+] = \dfrac{K_W}{[OH^-]} = \dfrac{K_W \cdot V}{h}$

$[H^+] = \dfrac{K_W V}{\sqrt{K_h V}} = \sqrt{K_W K_A V}$ since $K_h = K_W/(K_A.K_B)$

pH $= -\frac{1}{2} \log K_W - \frac{1}{2} \log K_A - \frac{1}{2} \log V$
$= 7 + \frac{1}{2}pK - \frac{1}{2} \log V$.

solution of 0.05M salt
weak acid/ weak alkali

0.05M NH_4CN

K_A for HCN = 7.2×10^{-10} ;
K_B for NH_4OH = 1.75×10^{-5}
pH = $-\frac{1}{2} \log K_W - \frac{1}{2} \log K_A$
$\qquad + \frac{1}{2} \log K_B$
= 7 + 4.571 − 2.378
= 9.19 [irrespective of
concentration]

solubility and solubility
product of AgCl

---

solubility of AgCl = $1.46 \times$
$\qquad\qquad 10^{-4}$g/100cm^3

= 1.46×10^{-3}g/litre

= $\dfrac{1.46 \times 10^{-3}}{143.5}$ mol/litre

$\approx 1 \times 10^{-5}$ mol/litre

$K_s = 2 \times 10^{-10}$
= $[Ag^+] \times [Cl^-]$

$[Ag^+] = [Cl^-]$ in water

solubility = $[Ag^+] = \sqrt{K_s}$

= $\sqrt{2 \times 10^{-10}}$

= 1.41×10^{-5}

For a salt of weak alkali/strong acid:

$[H^+] = h/V$ and $h = \sqrt{K_h.V}$

$[H^+] = \sqrt{\dfrac{K_W.V}{K_B}} \times \dfrac{1}{V} = \sqrt{\dfrac{K_W}{K_B.V}}$

pH = $-\frac{1}{2} \log K_W + \frac{1}{2} \log K_B + \frac{1}{2} \log V$

$\quad = \frac{1}{2}pK + \frac{1}{2} \log V$ ($\log K_A + \log K_B = \log K_W$).

For a salt of weak alkali/weak acid:

$K_h = \dfrac{[h/V] \times [h/V]}{[1-h]/V \times [1-h]/V} = \dfrac{h^2}{(1-h^2)} \approx h^2$

i.e. K_h is independent of dilution

$K_h = \dfrac{K_W}{K_A \times K_B}$ therefore $h = \sqrt{\dfrac{K_W}{K_A.K_B}}$

$[H^+] = \sqrt{\dfrac{K_W.K_A}{K_B}}$

pH = $-\frac{1}{2} \log K_W - \frac{1}{2} \log K_A + \frac{1}{2} \log K_B$
pH = 7 + $\frac{1}{2}pK^A - \frac{1}{2}pK^B$ where these are pK values
for acid and base.

solubility product in a saturated solution containing
excess undissolved solid electrolyte, there is an
equilibrium between the excess solid and the ions in
solution: $B_x A_y \rightleftharpoons xB^+ + yA^-$. The equilibrium
constant, from the law of mass action (p.181) is:

$$K = \dfrac{[B^+]^x.[A^-]^y}{[B_xA_y]}$$

By convention, the concentration of the solid electrolyte
is taken as unity, so: $K_s = [B^+]^x[A^-]^y$, where K_s is the
solubility product. The solubility product is constant
irrespective of other electrolytes present in the solution.
For sparingly soluble electrolytes, with no other added
electrolyte, the activity coefficients of the ions will not
differ from unity, so the concentration of the ion can be
used. For univalent sparingly soluble electrolytes,
$K_s = [B^+][A^-]$. The solubility product, K_s, is related to
the solubility in pure water, S_o, of sparingly soluble
electrolytes. Let the solubility be S_o mol/litre, then
$[B^+] = xS_o$ and $[A^-] = yS_o$, from the equation above.

$$K_s = [xS_o]^x.[yS_o]^y = [S_o]^{x+y}.x^x.y^y.$$

For a uniunivalent electrolyte, $K_s = S_o^2$.

common ion effect the solubility of any salt is less in a solution containing a common ion than in pure water, providing there is no formation of complex ions (↓). e.g. the solubility of AgCl (a sparingly soluble substance) in a solution of KCl is less than in pure water. The value of [Cl⁻] is determined by the total amount of chloride ions, from KCl and AgCl. Since [Ag⁺].[Cl⁻], the solubility product, is constant, [Ag⁺] must be less in the solution, and it is a measure of the solubility of AgCl. For general conditions of sparingly soluble electrolytes, activity coefficients have to be applied to all concentrations. With a common ion, solubilities are decreased with increasing concentration of the common ion. With no common ion added, solubilities increase with an increasing amount of electrolyte added. The behaviour of a slightly soluble electrolyte in varied concentrations of an added electrolyte is shown in the diagram.

complex ion a term generally reserved for ions which consist of a central metallic cation to which a number of ions or molecules are bonded, mostly by coordinate bonds (p.77). The ions or molecules are referred to as ligands (p.78). Water is included amongst ligands, so a hydrated ion may be a complex ion. In a solution, a ligand and a solvent molecule may compete for a coordination position, e.g.
$[Ag(H_2O)_2]^+ + 2NH_3 \leftrightarrows [Ag(NH_3)_2]^+ + 2H_2O.$

complex cation a complex ion (↑) in which the overall charge from the metallic ion and its ligands remains positive. It is named from the number and names of the ligands, the metallic ion, and the oxidation number of the metallic ion.

complex anion a complex ion (↑) in which the overall charge from the metallic ion and its ligands is negative. It is named from the number and name of the ligands, the central metallic ion ending in -ate, and the oxidation number of the metallic ion.

instability constant the dissociation constant for a complex ion, e.g. for the reaction:
$[Ag(NH_3)_2]^+ + 2H_2O \leftrightarrows [Ag(H_2O)_2]^+ + 2NH_3$
$$\text{instability constant} = K = \frac{[Ag(H_2O)_2]^+ \, [NH_3]^2}{[Ag(NH_3)_2]^+}$$
with water concentration taken as unity. The higher the instability constant, the less stable the complex ion.

formation constant the reciprocal of the instability constant (↑), using activities instead of concentrations.

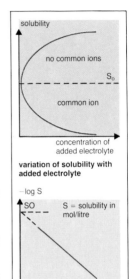

variation of solubility with added electrolyte

determination of K_s in solutions with no common ion

$[Ag(NH_3)_2]^+$
diammine silver(I)

$[Cr(Cl)_2(H_2O)_4]^+$
dichlorotetraaquo chromium (III)

$[Cu(NH_3)_4]^{2+}$
tetraammine copper (II)
complex cations

$[Fe(CN)_5(CO)]^{3-}$
pentacyanocarbonyl ferrate (II)

$[Cr(Cl)_4(H_2O)_2]^-$
tetrachlorodiaquo chromate (III)

$[Pt(Cl)_6]^{2-}$
hexachloroplatinate (IV)

$[Zn(OH)_4]^{2-}$
tetrahydroxozincate (II)
complex anions

stability constant same as **formation constant** (↑).
kinetic theory a theory which treats matter as consisting
of particles in motion; it is applicable mainly to gases,
but can be used to explain some of the properties of
liquids and solids. The principles of mechanics are
applied to the particles. The theory has the following
postulates: (1) matter consists of particles that can be
molecules, atoms, or ions. (2) in gases, the particles are
separated by distances that are large in comparison
with their size. In liquids the distances are much
smaller. (3) the particles are in continuous random
motion which is random in gases and liquids. (4) the
particles in gases and liquids collide with each other
and all collisions are perfectly elastic. (5) the particles
possess kinetic energy, which is a measure of the
temperature of matter; the average kinetic energy of all
particles in a particular specimen of matter is
considered constant for a given temperature. (6) the
pressure of a gas is considered to arise from the
bombardment of the walls of a containing vessel by the
particles of the gas.
molecular velocity in a gas each molecule has its own
velocity, symbol u.
Maxwell's distribution law the distribution of velocities
of molecules of a gas is calculated from the laws of
probability and the continual interchange of momentum
between molecules because of their frequent collisions.
The distribution of velocities depends on temperature.
law of distribution of molecular velocities an
alternative term for **Maxwell's distribution law** (↑).
mean velocity if the velocities of individual molecules
are u_1, u_2, u_3, etc. then the mean (or average) velocity,
$\bar{u}$, is given by:
$$\bar{u} = \frac{u_1 + u_2 + u_3 + u_4 + \dots}{n}$$
where n is the number of molecules. $\bar{u} = \sqrt{8RT/\Pi M}$,
where R is the molar gas constant (p.148), T the
absolute temperature (p.108) and M is the molar mass.
mean square velocity if the velocities of individual
molecules are u_1, u_2, u_3, etc. then the mean square
velocity, $\bar{u}^2$, is:
$$\bar{u}^2 = \frac{u_1^2 + u_2^2 + u_3^2 + \dots}{n}$$
where n is the number of molecules. $\bar{u}^2 = 3RT/M$ where
R is the molar gas constant (p.148), T the absolute temperature (p.108) and M the molar mass (p.29) of the gas.

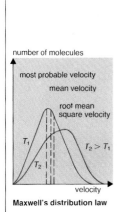

number of molecules

most probable velocity

mean velocity

root mean square velocity

T_1 $T_2 > T_1$ T_2

velocity

Maxwell's distribution law

mean velocity
oxygen molecules at 25°C

$R = 8.31\,\mathrm{Jk^{-1}\,mol^{-1}}$

$T = 298\,\mathrm{K}$

$M = 32\,\mathrm{g} = 0.032\,\mathrm{kg}$

$\bar{U} = \sqrt{\dfrac{8RT}{\pi M}}$

$= \sqrt{\dfrac{8 \times 8.31 \times 298}{3.1416 \times 0.032}}$

$= 444\,\mathrm{ms^{-1}}$

root mean square velocity if $\bar{u}^2$ is the mean square velocity, then $\bar{u}_{rms} = \sqrt{\bar{u}^2}$ is the root mean square velocity. $\bar{u}_{rms} = \sqrt{3p/\rho}$, where p is the pressure and ρ the density of the gas at s.t.p.

most probable velocity the maximum of the curve for Maxwell's distribution law (p.143) gives the most probable velocity, which is the velocity possessed by more molecules than any other velocity. If u is the most probable velocity, then: $u = \sqrt{2RT/M}$, where R is the molar gas constant (p.148), T the absolute temperature and M the molar mass (p.29) of the gas.

kinetic equation based on the kinetic theory (p.145), a general kinetic equation for gases is: $pV = \frac{1}{3} nm\bar{u}^2$ where p is the pressure exerted by a gas in a volume V, m is the mass of one molecule, n the number of molecules, and $\bar{u}^2$ the mean square velocity.

kinetic energy for a molecule of mass m and velocity u, the kinetic energy of the molecule is $\frac{1}{2} mu^2$.

energy distribution the distribution of kinetic energy of molecules in gases follows the same pattern as the law of distribution of molecular velocities (p.143). Expressed mathematically, if n_o is the total number of molecules, then the number of molecules, n, having a kinetic energy greater than the value E is:
$$n = n_o e^{-E/RT}$$
where R is the molar gas constant (p.148) and T the absolute temperature (p.108).

collision diameter the distance between the centres of two colliding molecules at the point of closest approach is the collision diameter; its symbol is σ.

mean free path the mean distance a molecule travels before it collides with another molecule. If l is the mean free path, η the coefficient of viscosity (p.155), p the pressure of a gas, and ρ its density, then:
$$l = \eta \sqrt{3/p\rho}$$

collision number if $\bar{u}$ is the mean velocity (p.143) of molecules in a gas and l is the mean free path, then the number of collisions per second made by a molecule is $\bar{u}/l$. The collision number, Z, is given by:
$Z = 2n^2\sigma^2 \sqrt{2RT/M}$, where n is the number of molecules in unit volume and σ is the collision diameter ($\uparrow$). Z is the number of collisions per unit volume.

Boltzmann factor the exponential factor in the equation which determines the fraction of molecules with a kinetic energy greater than a certain value, *see* **energy distribution** ($\uparrow$). The factor is $e^{-E/RT}$.

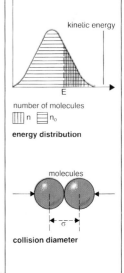

root mean square velocity
oxygen molecules at 25°C

$R = 8.31$ Jk^{-1} mol^{-1}
$T = 298$K
$M = 32$g $= 0.032$kg

$$\sqrt{\bar{u}^2} = \sqrt{\frac{3RT}{M}}$$

$$= \sqrt{\frac{3 \times 8.31 \times 298}{0.032}}$$

$$= 482 \text{ms}^{-1}$$

fraction of molecules

kinetic energy

E

number of molecules
⫿ n ⊟ n_o

energy distribution

molecules

σ

collision diameter

Boltzmann equation in a given volume there are n_1 molecules in state 1, and n_2 molecules in state 2. The difference in potential energy between states 1 and 2 is ΔE. If both states are at the same temperature then: $n_2/n_1 = e^{-\Delta E\,RT}$ (Boltzmann equation), where R is the molar gas constant and T the absolute temperature. This is a general equation with useful applications in physical chemistry.

Boyle's law the volume of a given mass of gas is inversely proportional to its pressure, provided the temperature remains constant. Expressed mathematically, $pV = kT$, where p is the pressure, V the volume, and T the absolute temperature of the given mass of gas, k is a constant for the given mass of gas. Real gases obey the law at low pressures; at high pressures they deviate from the law. *See* **ideal gas** (↓).

graphs for Boyle's law

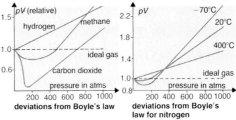

deviations from Boyle's law | deviations from Boyle's law for nitrogen

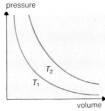

isobars for an ideal gas

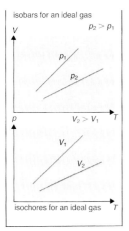

isochores for an ideal gas

ideal gas a gas which obeys Boyle's law (↑) and Gay-Lussac's law (↓) at all temperatures and pressures.

perfect gas the same as an **ideal gas** (↑).

non-ideal gas for a non-ideal gas the product of pressure and volume does not remain constant as pressure is increased. Deviations from Boyle's law depend on temperature and pressure and vary for different gases.

Gay-Lussac's law the volume of a given mass of gas changes by the same fraction (1/273) of its volume at 0°C for every degree rise in temperature, provided the pressure remains constant.

Charles' law same as **Gay-Lussac's law** (↑).

isobar[2] (n) an alternative statement of Gay-Lussac's law (↑) is: the volume of a given mass of gas is directly proportional to the absolute temperature, provided the pressure remains constant. A graph showing that $V/T =$ constant is an isobar.

isochore (n) a graph showing the relation of pressure and temperature for a constant volume of a given mass of a gas. The graph shows that $p/T =$ constant.

absolute zero all gases approach ideal conditions at low pressures. By extrapolating to zero volume of a gas at constant pressure it is found that the volume decrease is 1/273.15 of the volume at 0°C. The temperature at zero volume is absolute zero and this is −273.15°C.

kelvin (n) to measure absolute temperatures, absolute zero (↑) is taken as 0 degrees and the temperature interval, the kelvin, is the fraction 1/273.16 of the triple point (p.160) of water. This makes absolute zero −273.15°C, i.e. 0°C = 273.15 K. The approximate value of 273° is sufficient for most calculations so t°C = (t + 273)K. The symbol for kelvin is K, without the sign (°). Absolute temperatures are measured in kelvin.

equation of state by combining Boyle's law and Gay-Lussac's law (↑), the equation of state is derived, i.e. pV/T = constant. This is usually expressed as:

$$\frac{p_1 V_1}{T_1} = \frac{p_2 V_2}{T_2} = \frac{p_3 V_3}{T_3} = \ldots$$

is for a given mass of gas; it is obeyed by an ideal gas (p.145).

gas equation by extension of the equation of state (↑): $pV = nRT$ is derived, where p is the pressure, V the volume, T the absolute temperature, n the number of moles of gas, and R is the molar gas constant. The equation is true for an ideal gas.

ideal gas law equation the same as the **gas equation** (↑).

molar gas constant the constant defined by $R = pV/nT$ from the gas equation (↑). Its units are calculated from:

$$R = \frac{\text{pressure} \times (\text{length})^3}{\text{degrees} \times \text{moles}} = \frac{\text{force} \times (\text{length})^{-2} \times (\text{length})^3}{\text{degrees} \times \text{moles}}$$

$$= \frac{\text{force} \times \text{length}}{\text{degrees} \times \text{moles}} = \frac{\text{energy}}{\text{degrees} \times \text{moles}} = JK^{-1}mol^{-1}$$

$R = 8.314\,J\,K^{-1}\,mol^{-1}$ or
$R = 0.082054$ litre-atm K^{-1} mol^{-1}.

gas constant see **molar gas constant** (↑). It can also be represented by r, where $r = nR$ for a particular mass of gas.

Boltzmann constant a constant defined from $k = R/L$, where R is the molar gas constant (↑) and L is the Avogadro constant (p.30). The units are J K^{-1}, and $k = 1.380 \times 10^{-23}$ J K^{-1}. It can be regarded as the gas constant per single molecule.

Avogadro's hypothesis equal volumes of all gases under the same conditions of temperature and pressure contain the same numbers of molecules.

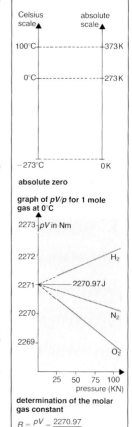

absolute zero

graph of pV/p for 1 mole gas at 0°C

determination of the molar gas constant

$$R = \frac{pV}{T} = \frac{2270.97}{273.15}$$

$$= 8.314\,JK^{-1}mol^{-1}.$$

data extrapolated to zero pressure

$pV = 2270.97\,J$

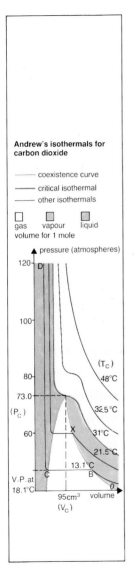

Andrew's isothermals for carbon dioxide

——— coexistence curve

——— critical isothermal

——— other isothermals

☐ ▨ ▨
gas vapour liquid
volume for 1 mole

Avogadro's law Avogadro's hypothesis (↑) now accepted as law.

Joule-Thomson effect a compressed gas, at 250 atm pressure, is allowed to stream through a porous plug or a throttle and, as a result, the pressure falls, the gas expands and energy is needed to overcome intermolecular attraction. Energy is taken from the kinetic energy of the molecules, their speed decreases, and the gas is cooled. *See* **inversion temperature** (↓).

inversion temperature at room temperatures, some gases, e.g. hydrogen and helium, have a negative Joule-Thomson effect (↑), and the temperature rises when the gas is expanded. Most gases have a value of the product pV which is less at a higher pressure, and cooling results from this. *See* **non-ideal gases** (p.147). If the value of pV increases with rising temperature, there is a negative Joule-Thomson effect. The inversion temperature is that above which there is no cooling from the Joule-Thomson effect. All gases have an inversion temperature; for helium it is 25 K, for nitrogen it is 650 K.

van der Waals' equation the attractive force between the molecules of a non-ideal gas reduces the pressure of the molecules on the walls of the containing vessel. If p is the actual pressure, then the ideal pressure is $p + a/V^2$ where a is a constant for a particular gas. The molecules occupy an appreciable volume, so the ideal volume is $V - b$, where b is a constant for a particular gas. The product of ideal pressure and ideal volume is given by van der Waals' equation for a gas: $(p + a/V^2)(V - b) = nRT$. This gives better agreement with experimental values than the gas equation (↑).

isothermals (*n.pl.*) lines connecting points of equal temperature.

Andrews' isothermals a set of isothermals (↑) for carbon dioxide, with values for one mole of substance as shown. At 48°C a normal p, V isothermal for a gas is obtained. At 32.5°C a large deviation from a normal isothermal is obtained. At 31°C the gas is liquefied at a pressure of 73.0 atmospheres. At 13.1°C, the portion of the curve, AB, is the isothermal for a vapour; along the portion, BC, vapour and liquid can coexist. The portion, CD, is for a liquid, and the steepness of the curve shows how incompressible a liquid is. The portion of the curve, BC, is at a pressure corresponding to the vapour pressure at that isothermal temperature. The point, X, represents carbon dioxide in its critical state.

critical isothermal the pressure-volume curve at the
critical temperature ($\downarrow$).

critical temperature the maximum temperature at which
a gas can be liquefied (p.154). Above this temperature,
a liquid state of a substance cannot exist. Symbol T.

substance	T_c (K)	P_c (atm)	molar volume
helium	5·2	2·26	60 cm^3
hydrogen	33·2	12·8	68 cm^3
nitrogen	126·0	33·5	90 cm^3
oxygen	154·3	49·7	74 cm^3
carbon dioxide	304·2	73·0	95 cm^3
ammonia	406·0	112·3	72 cm^3
benzene	561·6	47·9	256 cm^3
water	647·3	217·7	57 cm^3

critical pressure the pressure which causes liquefaction
(p.154) at the critical temperature ($\uparrow$). Symbol p_c.

critical volume the volume occupied by a given mass of
gas at its critical pressure and temperature. If the
amount of gas is one mole, this is the molar critical
volume. The molar critical volume is usually quoted for a
gas. Symbol: V_c.

critical density the density of a gas at its critical
pressure and temperature. Symbol ρ_c. The density of
gas and liquid are identical at the critical density.

critical constants the critical temperature T_c, the critical
pressure p_c, the critical density ρ_c and the molar critical
volume V_c.

critical point the point, X, on the critical isothermal ($\uparrow$),
i.e. where a gas is at its critical pressure and
temperature.

critical state the state at the critical point ($\uparrow$).

coexistence curve the curve marking the boundary of
the coexistence of liquid and vapour in a set of
pressure-volume isothermals.

vapour (n) a gaseous substance when its temperature is
below the critical temperature ($\uparrow$). A vapour can be
liquefied by pressure alone, a gas cannot be liquefied
by pressure alone. **vaporize** (v).

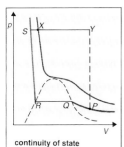

continuity of state

compressibility curves

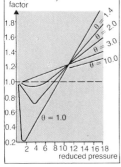

continuity of state at the critical point, Z, *see diagram*, the densities of the gaseous and the liquid states of a substance are identical and the surface of separation between the two phases disappears. A gas at point P in the diagram is compressed at constant volume to the point Y; the temperature is raised. The gas is then cooled at constant pressure to point S. The volume decreases, and at the point X on the critical isothermal (↑), the gas changes to a liquid without any discontinuity of state. Along the line YS the change from gas to liquid is continuous, exhibiting continuity of state.

permanent gas a gas whose critical temperature (↑) is below room temperature. The term was in use when it was impossible to liquefy gases by pressure alone.

reduced pressure the pressure of a gas expressed as a fraction of the critical pressure (↑); its symbol is Π. If p is the pressure and p_c the critical pressure of a gas, then $\Pi = p/p_c$.

reduced volume the volume of a gas expressed as a fraction of the critical volume (↑); its symbol is Φ. If V is the volume and V_c the critical volume, then $\Phi = V/V_c$.

reduced temperature the temperature of a gas expressed as a fraction of the critical temperature (↑); its symbol is θ. If T is the temperature, T_c the critical temperature of a gas, then $\theta = T/T_c$.

reduced equation of state the reduced pressure, volume and temperature of a gas (↑) are substituted in van der Waals' equation (p.149), giving:

$$\left(\pi p_c + \frac{a}{\Phi^2 V_c^2}\right)\ (\Phi V_c - b\ ) = R\theta T_c$$

van der Waals' equation can be solved for the critical state giving the results of: $p_c = a/27b^2$; $V_c = 3b$; $T_c = 8a/27Rb$. These results are substituted in the equation above, forming the new equation:

$$\left(\pi + \frac{3}{\Phi^2}\right)\ (3\Phi - 1) = 8\theta$$

which is the reduced equation of state.

corresponding states if two or more substances have the same values for reduced pressure, temperature and volume (↑), then they are in corresponding states.

compressibility factor the deviation of a gas from ideal behaviour is defined and expressed by the compressibility factor, K, where $K = pV/RT$ (values for 1 mole of gas). For an ideal gas, $K = 1$; for a non-ideal gas, K is greater or less than 1.

Boyle temperature the temperature at which a gas most closely obeys Boyle's law.

vapour density the ratio of the mass of a given volume of gas to the mass of the same volume of hydrogen, both volumes are measured at the same temperature and pressure. It can also be measured from:

$$\text{vapour density} = \frac{\text{density of gas}}{\text{density of hydrogen}}$$

with both densities measured at the same pressure and temperature. The vapour density of a gas is numerically equal to half its relative molecular mass. The symbol for vapour density is Δ.

thermal dissociation if a gas, liquid or solid splits up reversibly on heating to form other molecules or atoms, it undergoes *thermal dissociation*. Examples are: $PCl_5 \rightleftharpoons PCl_3 + Cl_2$; $NH_4Cl \rightleftharpoons NH_3 + HCl$; $N_2O_4 \rightleftharpoons 2NO_2$.

degree of dissociation if thermal dissociation (↑) leads to an increase in the number of molecules, the vapour density (↑) of a gas is changed. In the reaction:

$$NH_4Cl \rightleftharpoons NH_3 + HCl$$
$$(1 - u)\text{moles} \quad u\text{ moles} \quad u\text{ moles}.$$

where u is the degree of dissociation, the volume change results in:

$$\frac{\text{volume if dissociated}}{\text{volume no dissociation}} = \frac{1 + u}{1} = \frac{\Delta_1 \text{ no dissociation}}{\Delta_2 \text{ if dissociated}}$$

whence $u = \dfrac{\Delta_1 - \Delta_2}{\Delta_2}$ where Δ is the vapour density.

association (n) substances which have relative molecular masses larger than those expected from simple molecules exhibit molecular association, e.g., $2CH_3COOH \rightleftharpoons (CH_3COOH)_2$ the molecules of ethanoic acid undergo association to form double molecules.

law of partial pressures the partial pressure of each gas in a mixture, providing there is no reaction between the gases, is equal to the pressure the gas would exert if it alone occupied the whole volume of the mixture at the same temperature. The law of partial pressures states : the total pressure of a mixture of gases is equal to the sum of the partial pressures of each gas in the mixture. Expressed mathematically:

$$p = p_1 + p_2 + p_3 + \ldots$$

Dalton's law the law of partial pressures (↑).

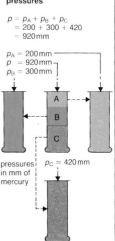

Dalton's law of partial pressures

$p = p_A + p_B + p_C$
$= 200 + 300 + 420$
$= 920\text{mm}$

$p_A = 200\text{mm}$
$p = 920\text{mm}$
$p_B = 300\text{mm}$

$p_C = 420\text{mm}$

pressures in mm of mercury

diffusion (*n*) the process by which any substance spreads uniformly throughout the space available to it. The process in gases is extremely quick, in solids it is infinitesimally slow. Diffusion is also the process by which gases pass through porous materials when the pressure is equal on both sides of the porous material. Contrast **effusion** (↓). **diffuse** (*v*).

effusion (*n*) the passage of a gas through a fine orifice or a porous material from a region of higher pressure to one of lower pressure. Contrast **diffusion** (↑). **effuse** (*v*).

Graham's law of diffusion the rate of diffusion of a gas is inversely proportional to the square root of its density; it is an approximate law only. Two rates of diffusion, D_1 and D_2, can be compared at the same temperature:

$$\frac{D_1}{D_2} = \sqrt{\frac{\rho_2}{\rho_1}}$$

where ρ is the density of a gas. The density of a gas is proportional to its relative molecular mass, M, so:

$$\frac{D_1}{D_2} = \sqrt{\frac{M_2}{M_1}}$$

The law applies to the effusion (↑) of gases as well; comparing the effusion of two gases requires the same conditions of temperature and pressure.

It takes 165 s for a volume of oxygen and 251 s for the same volume of the vapour of a substance to effuse through an orifice under the same conditions. Calculate the molar mass of the substance.

$$\frac{D_1}{D_2} = \sqrt{\frac{M_2}{M_1}} = \frac{t_2}{t_1}$$

$t_2 = 251$ s; $t_1 = 165$ s; $M_1 = 32$g

$$M_2 = \frac{t_2^2 \times M_1}{t_1^2} = \frac{(251)^2 \times 32}{(165)^2} \text{ g}$$

$$= 74\text{g}$$

boiling-point relationship the boiling point of a liquid, T_b, at 1 atmosphere pressure is approximately two-thirds of the critical temperature, T_c, both temperatures being measured on the absolute scale. $T_b/T_c = 2/3$.

liquefaction of gases the process by which a gas
becomes a liquid; it is brought about by pressure and
by cooling. A gas has to be below its critical
temperature (p.150) before it can be liquefied. Cooling
is effected by the Joule-Thomson effect (p.149), or by
adiabatic expansion (↓). *See* **continuity of state** (p.151).

adiabatic expansion² a gas is compressed to about 200
atmospheres, then allowed to expand in an engine
where it does mechanical work. Kinetic energy is removed from the gas molecules to perform the work and
the gas is cooled. *See* **adiabatic expansion¹** (p.114).

condensation (*n*) the process in which a vapour
changes to a liquid. Heat is given out during the
process. Condensation occurs when a vapour is
cooled. Contrast **liquefaction** (↑).

vaporize (*v*) to cause a liquid to change to a vapour by
raising its temperature, by decreasing the external
pressure, or mechanically by blowing air through it.

evaporate (*v*) to change a liquid to vapour at any
temperature.

molar heat of vaporization the quantity of heat to
vaporize 1 mole of a substance.

vapour pressure the pressure at which liquid and
vapour can coexist in equilibrium; the rate of
evaporation is equal to the rate of condensation (↑).
Vapour pressure increases with temperature. The
critical pressure (p.150) is the highest possible vapour
pressure of a liquid substance. Vapour pressure is
independent of the amount of liquid present, and the
vapour pressure/temperature curves are similar for all
liquids. A liquid boils when its vapour pressure is equal
to the atmospheric pressure, or other external pressure.
See **Clausius-Clapeyron equation** (p.115).

saturated vapour pressure the correct term for **vapour
pressure** (↑).

surface tension a molecule on the surface of a liquid
has a net cohesive force acting inwards, as there are
more molecules per unit volume in the liquid than in its
vapour. There is thus a resultant force acting on the
liquid surface so that the surface tends to contract to
the smallest area. For a drop, the smallest possible area
for a given volume is a sphere, causing small drops to
be spherical. This surface force is called surface
tension, as the surface behaves as if it were in a state of
tension. Surface tension is defined as: the tangential
force acting at right angles in the surface of a liquid to a

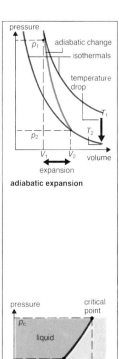

adiabatic expansion

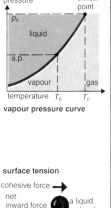

vapour pressure curve

surface tension

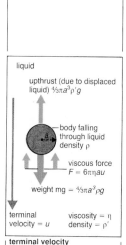

liquid

upthrust (due to displaced liquid) $\frac{4}{3}\pi a^3 \rho' g$

body falling through liquid density ρ

viscous force $F = 6\pi\eta au$

weight $mg = \frac{4}{3}\pi a^3 \rho g$

terminal velocity = u viscosity = η density = ρ'

terminal velocity

molar heat capacities

gas	He	Hg	I vapour
C_p (J)	20·9	20·9	20·9
C_v (J)	12·6	12·5	12·5
C_p-C_v (J)	8·3	8·4	8·4
γ	1·66	1·67	1·67

line of unit length in the surface of the liquid. Surface tension usually decreases with an increase in temperature and falls to zero at the critical temperature of a substance. The symbol for surface tension is γ and it is measured in newtons per metre. For water, $\gamma = 7.28 \times 10^{-2}\,\mathrm{N\,m^{-1}}$ at 293 K.

coefficient of viscosity the force per unit area required to maintain a difference of velocity of $1\,\mathrm{m\,s^{-1}}$ between two parallel layers of liquid 1 m apart. The symbol for the coefficient of viscosity is η and its units are newton seconds per metre squared ($\mathrm{Ns\,m^{-2}}$).

Stokes' law the viscous friction, F, acting on a sphere of radius, a, falling through a fluid of viscosity η with a velocity v is given by: $F = 6\pi\eta av$. *See* **terminal velocity.**

terminal velocity a body falling through a fluid reaches a constant velocity, its terminal velocity, when the downward gravitational force is balanced by the upward viscous friction, F, and the upthrust of the displaced fluid. For a sphere of radius, a, and density, ρ, falling with a terminal velocity, u, through a fluid of viscosity η and density ρ', and g as the acceleration due to gravity:

F = weight of sphere − upthrust on sphere

$$6\pi\eta au = \frac{4}{3}\pi a^3\rho g - \frac{4}{3}\pi a^3\rho' g$$

whence

$$u = \frac{2}{9}\frac{ga^2(\rho - \rho')}{\eta}$$

(also called Stokes' law (↑))

heat capacity at constant volume the quantity of heat required to raise the temperature of a gaseous system at constant volume by one degree kelvin.

specific heat capacity at constant volume the heat capacity at constant volume (↑) of 1 kg of gas; its symbol is c_v; units are $\mathrm{J\,kg^{-1}\,K^{-1}}$.

molar heat capacity at constant volume the heat capacity at constant volume (↑) of one mole of gas; its symbol is C_v; units are $\mathrm{J\,mol^{-1}\,K^{-1}}$. *See* **translational energy** (p.156).

heat capacity at constant pressure the quantity of heat required to raise the temperature of a gaseous system at constant pressure by one degree kelvin.

specific heat capacity at constant pressure the heat capacity at constant pressure (↑) of 1 kg of gas; its symbol is c_p; units are $\mathrm{J\,kg^{-1}\,K^{-1}}$.

molar heat capacity at constant pressure the heat capacity at constant pressure (↑) of one mole of gas; its symbol is C_p; units are $\mathrm{J\,mol^{-1}\,K^{-1}}$.

translational energy the energy of gaseous molecules arising from their kinetic energy of motion. If the temperature of a gas is raised, at constant volume, the heat supplied increases the kinetic energy of the molecules alone. The kinetic energy of all the molecules in one mole of gas, E_K, is derived from the gas equation:

$$pV = \tfrac{1}{3}Lmu^2 = \tfrac{2}{3}.L.\tfrac{1}{2}mc^2 = \tfrac{2}{3}E_K = RT.$$

So the increase in kinetic energy per degree
= $\tfrac{3}{2}R(T+1) - \tfrac{3}{2}RT = \tfrac{3}{2}R$, and this is C_v;
$C_v = \tfrac{3}{2} \times 8.3143\,\mathrm{J\,K^{-1}\,mol^{-1}} = 12.471\,\mathrm{J\,K^{-1}\,mol^{-1}}$

rotational energy energy due to the rotation of a molecule about the three axes of space; the molecule has two or more atoms.

vibrational energy energy due to oscillation of the atoms within a molecule; the molecule has two or more atoms.

principle of the equipartition of energy each form of energy, translational (↑), rotational (↑) and vibrational (↑) manifested by the molecules in one mole of gas contributes $\tfrac{1}{2}R$ to the molar heat capacity (p.155). R is the molar gas constant.

non-ideal gases

ideal gases

	γ
monatomic	1·67
diatomic	1·40
triatomic	1·33

gas	C_p (J)	C_γ (J)	C_p-C_γ (J)	γ
chlorine	34·1	25·1	9·0	1·36
hydrogen chloride	29·6	21·1	8·6	1·41
sulphur dioxide	40·6	31·4	9·2	1·29
ammonia	37·2	28·4	8·8	1·31

ratio of molar heat capacities the value of C_p is greater than C_v as work is done in expansion. If the volume of one mole of gas increases from V_1 to V_2, then $pV_1 = RT$ $pV_2 = R(T+1)$. Work done = $p(V_2 - V_1)$ = R = 8.3143 J. The ratio C_p/C_v for a monatomic gas, i.e. with no rotational or vibrational energy (↑) is:

$$\frac{C_p}{C_v} = \frac{12.471 + 8.3143}{12.471} = 1.67$$

For a polyatomic gas, C_p is greater than for a monatomic gas. *See table* for ratios of particular gases. The symbol for the ratio of molar heat capacities is γ.

atomicity (n) the number of atoms in a molecule of an element or a compound, e.g. the atomicity of hydrogen is 2 (H_2); the atomicity of sulphur dioxide is 3 (SO_2); the atomicity of helium is 1 (He).

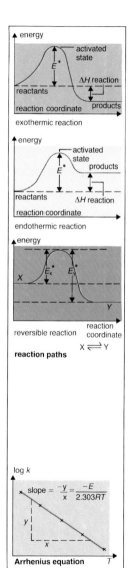

exothermic reaction

endothermic reaction

reversible reaction

reaction coordinate

$X \rightleftharpoons Y$

reaction paths

activated state

log k

slope $= \dfrac{-y}{x} = \dfrac{-E}{2.303RT}$

Arrhenius equation

reaction path a diagrammatic representation of a chemical reaction. It indicates the energy relationships of reactants, activated state (↓), products, activation energy (↓) and heat of reaction (p.103).

activation energy the additional energy the reactants must acquire to form an activated state (↓) for a reaction; it is the difference in energy between the activated state and the reactants. Energy is obtained from collisions between molecules, hence only a proportion of molecules with a sufficiently high speed to possess enough kinetic energy can supply sufficient energy to give a molecule its activation energy. If Y represents the total number of collisions between reactant molecules, then $Ye^{-E/RT}$ is the number of possible effective collisions to produce an activated state, where $E^{\ddagger}$ is the activation energy.

activated state an intermediate state in a chemical reaction. The reactant molecules approach, and with sufficient energy form an activated complex with appropriate rearrangement of valency bonds and energy. The state has a transient existence and then breaks up into the products.

activated complex *see* **activated state** (↑).

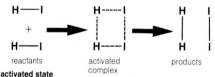

reactants   activated complex   products

Arrhenius equation an equation suggested from experimental evidence; it is $k = Ae^{-E\ddagger/RT}$ where k is the rate constant (p.181) for a reaction, A is a temperature-independent constant for a given reaction, sometimes called the **frequency factor**, $E^{\ddagger}$ is the activation energy (↑), R is the molar gas constant and T is the absolute temperature. Taking logarithms: $\log k = \log A - E^{\ddagger}/2.303RT$. By plotting $\log k$ against T, a value for $E^{\ddagger}$ is found from the slope of the graph.

steric term collision energy varies with the orientation of the reactant molecules. A molecule may have a reactive side and an unreactive side. A steric term has to be incorporated in the Arrhenius equation (↑) to account for a collision with sufficient energy not forming an activated complex (↑) because the orientation of the molecules was incorrect.

entropy of activation this determines the fraction of molecular collisions with a suitable orientation to produce an activated complex (p.157); the relation is:

$$\frac{\text{no. of appropriate orientations in collisions}}{\text{total no. of reactant collisions}} = e^{\Delta S^{\ddagger}/R}$$

where $\Delta S^{\ddagger}$ is the entropy of activation and R is the molar gas constant. $\Delta S^{\ddagger}$ decreases in value as the orientation requirements become stricter. This steric term (p.157) has to be incorporated in the Arrhenius equation (p.157) producing: $k = Ae^{-E^{\ddagger}/RT}e^{\Delta S^{\ddagger}/R}$

transition state theory a further description of the activated state (p.157) indicates the reversible formation of an activated complex (p.157), e.g. in the reaction

$$AB + C \rightarrow A + BC \text{ the stages are:}$$
$$AB + C \rightleftharpoons A...B...C \rightarrow A + BC$$

The intermediate product (A...B...C) can decompose to form either the original reactants or the final products. There may be more than one activated complex, with an intermediate product; the larger the activation energy, the slower the reaction, as fewer collisions result in the formation of an activated complex.

enthalpy of activation the Gibbs free energy, ΔG, the enthalpy, ΔH, the entropy, ΔS, the equilibrium constant, K, and the rate constant k, are related as follows:

$$\Delta G = -RT \ln K, \text{ so } K = e^{-\Delta G/R T}$$
$$\Delta G = \Delta H - T\Delta S, \text{ so substituting:}$$
$$K = e^{-\Delta H/RT}e^{\Delta S/R}$$

If the equilibrium constant is for the reaction to form an activated state (p.157) as in transition state theory (↑), then: $k = Ae^{\Delta S^{\ddagger}/R}e^{-\Delta H^{\ddagger}/RT}$, where $\Delta H^{\ddagger}$ is the enthalpy of activation, and A is the steric term.

exothermic reaction a reaction which gives out heat to its surroundings, the change in enthalpy, ΔH, is written as negative as the system loses heat. Exothermic reactions forming simpler molecules will always take place as the products are at a lower energy level and entropy has increased, e.g. combustion of fuels. Exothermic reactions in which more ordered products are formed, take place if the energy losses to other parts of the system are greater than the gain in entropy, e.g. precipitation reactions, condensation of vapours, crystallization and solidification.

endothermic reaction a reaction which takes in heat from its surroundings, the change in enthalpy, ΔH, is written as positive as the system gains heat.

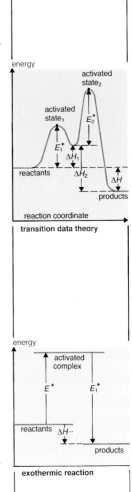

transition data theory

exothermic reaction

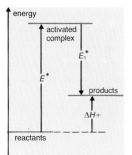

endothermic reaction

Endothermic reactions are less likely to occur than exothermic reactions because of the relationship $\Delta G = \Delta H - T\Delta S$. Any reaction which is endothermic and forms more ordered products is improbable, since both processes decrease entropy; a high-energy source has to be available in which the increase in entropy is greater than the decrease in the reaction, e.g. photosynthesis. Endothermic reactions forming simpler molecules take place if the increase in entropy is large enough to counterbalance the energy change required for the reaction, e.g. thermal decompositions, formation of a solution, evaporation.

phase (n) any homogenous (p.161) and physically distinct part of a chemical system which is separated from other parts of the system by a distinct boundary, is chemically different from other parts of the system, and can be separated mechanically from other parts of the system. A gas or a mixture of gases forms a phase as all gases are completely miscible. Miscible liquids form one phase, immiscible liquids form two or more phases depending on the number of immiscible liquids. Generally every solid forms a separate phase except in solid solutions.

component (n) an independent chemical constituent of a system by which the composition of a phase of the system can be defined. The components of a system are the least number of separate substances by which the composition of every possible phase can be defined, e.g. in the system:

$$MgCO_3 = MgO + CO_2,$$

there are two solid and one gaseous substances, but only two components as the third substance is not independent of the other two, but defined from them.

one component phase system

position on diagram	phases present	degrees of freedom
O (triple pt.)	solid, liquid, vapour	0
along OX	liquid, vapour	1
along OY	liquid, solid	1
along OZ	solid, vapour	1
area ZOX	vapour only	2
area YOX	liquid only	2
area ZOY	solid only	2

degree of freedom a variable factor such as temperature, pressure or concentration. The number of degrees of freedom is the smallest number that has to be fixed in order that a system can remain permanently in one position of equilibrium.

phase rule the number of degrees of freedom (↑), F, is related to the number of components (p.157), C, and the number of phases, P, by the relation:

$$F = C - P + 2.$$

phase diagram a diagram which shows the relationship of the phases of a system for two variable factors. A simple phase diagram is shown for one component, i.e. a pure substance, with variable factors of pressure and temperature. The curves indicate equilibrium conditions for any two phases. The curve, OX, is the vapour pressure curve of the liquid substance; any point on this curve has only one degree of freedom, i.e. if the temperature is fixed, then so is the vapour pressure. The curve, OY, gives the equilibrium conditions between solid and liquid phases and shows the change of melting point with pressure. The curve, OZ, is the vapour pressure curve of the solid substance. Above the critical temperature, T_c, the substance is gaseous, and cannot be condensed by a reduction in temperature.

triple point the point at which three phases are in equilibrium, the point O in the phase diagram. The phase diagram for the water system shows the triple point is at 0.0075°C and a pressure of 4.579mm. The freezing point of water is at 1 atmosphere pressure and 0°C. The critical temperature is 374°C and the critical pressure is 218 atmospheres. The melting point curve of ice. OC in the diagram, slopes to the left so that the melting point of ice decreases with increasing pressure, a phenomenon which differs from almost all other solid substances.

supercooling (n) a liquid can be cooled below its freezing point without solidifying, provided the liquid is free from dust particles, and is cooled in a clean smooth vessel; this is known as supercooling. The supercooled liquid has a vapour pressure curve, which is a continuation of the normal vapour pressure curve. The supercooled liquid is in a metastable (↓) equilibrium with its own vapour. The introduction of some solid, dust, etc. causes the supercooled liquid to freeze and the temperature to rise to the normal freezing point.

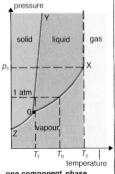

one component phase diagram

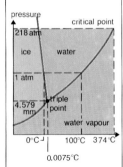

phase diagram for the water system

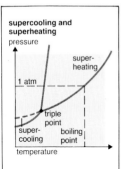

supercooling and superheating

pressure

super-heating

1 atm

triple point

super-cooling

boiling point

temperature

superheating (*n*) raising the boiling point of a liquid by increasing the pressure. Superheated steam is produced by heating water under pressure so that the boiling point is raised above 100°C. The surface tension in small bubbles formed when a liquid is heated in a perfectly clean, smooth vessel causes the vapour pressure to be raised sufficiently high to form such bubbles. This results in the temperature of the liquid being raised above the boiling point; this is also known as superheating. The result is boiling by bumping.

metastable (*adj*) describes a definite equilibrium, or a state, which is not the most stable at the given temperature, and needs only a very small impulse to revert to the normal equilibrium, or state, at that temperature, or a change to the temperature normally defined by the equilibrium or state, e.g. supercooled (↑) water is below the freezing point, but addition of a nucleus for crystallization causes the water to freeze and rise to the normal freezing point.

homogenous (*adj*) describes a state of matter that has the same composition and properties throughout the state, e.g. a solution is a homogenous mixture of solute and solvent with the same composition and properties throughout the solution; a homogenous system has all members in the same state of matter. **homogeneity** (*n*).

heterogenous (*adj*) describes a state of matter that does not have the same composition and characteristics throughout the state, e.g. the system $MgCO_3/MgO/CO_2$ is heterogenous as it contains substances in solid and gaseous states. **heterogeneity** (*n*).

phase equilibrium if a chemical system is in equilibrium at constant temperature and pressure, then the chemical potential (p.114) of any given component has the same value in every phase.

univariant (*adj*) a chemical system possessing one degree of freedom (↑).

bivariant (*adj*) a chemical system possessing two degrees of freedom (↑).

trivariant (*adj*) a chemical system possessing three degrees of freedom (↑).

invariant (*adj*) a chemical system possessing no degrees of freedom (↑).

condensed system a system in which only solid and liquid phases are considered; the effect of pressure is relatively small, so it is disregarded, and arbitrarily fixed at one atmosphere.

enantiotropic (*adj*) describes crystalline forms of a substance that can undergo a reversible change at a transition point (p.164). Phase diagrams give the relation between temperature and pressure, *see diagram*, for sulphur. B is the transition point at which rhombic changes to monoclinic sulphur; C is the melting point of monoclinic sulphur. BE is the metastable (p.161) vapour pressure curve for rhombic sulphur, and is followed if sulphur is heated rapidly, the solid turning to liquid at E. EC is the metastable vapour pressure curve for liquid sulphur. BF and CF show respectively the effect of pressure on the transition point and melting point of monoclinic sulphur. The area BCF gives the conditions under which monoclinic sulphur is stable.

azeotropic (*adj*) describes a mixture of miscible liquids which boils at a constant temperature. The composition as well as the boiling point of an azeotropic mixture changes with pressure. *see diagram*. If a liquid mixture represented by composition x_1 is distilled, the vapour has composition x_2 and condenses to form a liquid of that composition. Distillation starting at composition x_1 produces an azeotropic mixture, composition Z as the distillate, and the residue tends towards pure B. Similarly, a mixture of composition y_1 yields an azeotropic mixture as distillate, and pure A as a residue.

partial miscibility a phenomenon in which two liquids have limited solubility in each other. When the limited solubility is exceeded, the liquids form two immiscible layers; *see diagram* for the solubility curve for partial miscibility; such phase diagrams are for a constant pressure almost always atmospheric pressure.

critical solution temperature the maximum point of the solubility curve for partial miscibility (↑). Above this temperature, the two liquids are miscible in all proportions, and there is only one liquid phase.

conjugate solutions the points s_1 and s_2 in the solubility curve for partially miscible solutions show respectively a saturated solution of B in A and a saturated solution of A in B; these points give the compositions of the conjugate solutions. If a mixture of A and B is made, represented by the point X, then two immiscible liquid layers will be formed, and each layer will have the composition of a conjugate solution. As conjugate solutions are in equilibrium, they have the same vapour pressure.

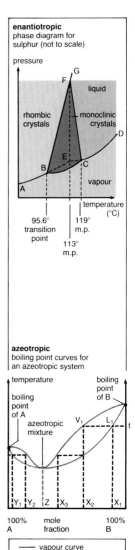

enantiotropic phase diagram for sulphur (not to scale)

pressure

rhombic crystals — monoclinic crystals

liquid

vapour

temperature (°C)

95.6° transition point

119° m.p.

113° m.p.

azeotropic boiling point curves for an azeotropic system

temperature

boiling point of A

boiling point of B

azeotropic mixture

100% A   mole fraction   100% B

—— vapour curve
—— liquid curve

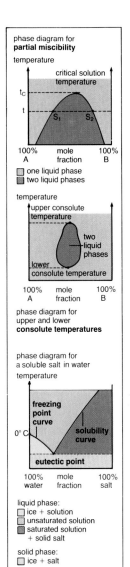

phase diagram for
partial miscibility

temperature

☐ one liquid phase
▨ two liquid phases

phase diagram for
upper and lower
consolute temperatures

phase diagram for
a soluble salt in water
temperature

liquid phase:
☐ ice + solution
☐ unsaturated solution
▨ saturated solution
  + solid salt

solid phase:
☐ ice + salt

consolute temperature a temperature at which two
partially miscible liquids become completely miscible.
The critical solution temperature (↑) is an upper
consolute temperature. Some partially miscible liquids
have a lower consolute temperature as well as an
upper. The solubility graph for partially miscible liquids
with both upper and lower consolute temperatures is
shown in the diagram.

freezing-point curve in a system with a liquid mixture in
equilibrium with a solid phase, and the solid phase is in
excess, the curve representing the variation of the
equilibrium temperature with composition is a freezing-
point curve.

solubility curve a curve representing the variation of the
equilibrium temperature of a solution and a solid phase
when the liquid phase is in excess.

solid solution the simplest solid solution has the
freezing point of all mixtures between the freezing
points of the pure components. The *liquidus*
curve gives the composition of the liquid phase
in equilibrium with the solid solution whose composition
is shown on the corresponding point of the *solidus*
curve. At a temperature t_1, a liquid of composition y will
be in equilibrium with a solid of composition x; these
compositions are determined by the tie line BC. If liquid
with a composition, and at a temperature indicated by
point A, is cooled, freezing will start at temperature t_1,
and the composition of the solid will be given by x, from
point C. As cooling proceeds to temperature t_2, the
composition of the liquid will change from y to z, and the
composition of the solid from x to y. Separation of solid
starts at B at a temperature t_1 and composition y; the
liquidus curve is thus the freezing-point curve (↑).
Melting of a solid of composition y starts at temperature
t_2; the solidus curve is thus the melting-point curve. For
a mixture of composition y, the freezing point is t_1 and
the melting point is t_2.

salt hydrates a salt can form more than one hydrate with
water, and the solubilities of each hydrate produce a
set of solubility curves. The phase diagrams for these
condensed systems can have either congruent or
incongruent melting points. In the curve of congruent
melting points, a saturated solution at that point can be
regarded as having the same composition as the solid
phase. The system for iron (III) chloride is shown with four
hydrates each with its own congruent melting point.

transition point an incongruent melting point, exhibited
by sodium sulphate which forms a decahydrate
($10H_2O$) and has solutions of the hydrated and
anhydrous forms. A metastable hydrate may also be
formed. Along the curve PQ, ice separates out on
cooling; Q is the eutectic point; along QR, Na_2SO_4.
$10H_2O$ separates out as the solution is saturated;
before a maximum is attained, a peritectic point is
reached, this is the transition point. At the transition
point the decahydrate decomposes into the anhydrous
salt. A metastable system may be formed with the
heptahydrate (Na_2SO_4. $7H_2O$).

salt hydrate equilibria if a salt forms more than one
hydrate, then a system of two hydrates with water
vapour is univariant (p.161) and at a given temperature
there is a fixed vapour pressure of the system. The
variation of vapour pressure for the copper (II) sulphate
system is shown in the diagram; the salt forms three
hydrates and has an anhydrous form. The vapour
pressure of the anhydrous form is 0mm. A graph of
vapour pressure against concentration shows the
equilibrium pressure. If the atmospheric water vapour
pressure is lower than the vapour pressure of a hydrate,
the hydrate loses water to the atmosphere and forms a
lower hydrate, or an anhydrous form. In the diagram, an
atmospheric water vapour pressure of 6mm causes
$CuSO_4.5H_2O$ to **effloresce** to $CuSO_4.3H_2O$. If the
reverse conditions of water vapour pressure exist then
the hydrate takes in water and turns into a saturated
solution, i.e. it **deliquesces**.

efflorescence (*n*) the phenomenon of efflorescing (↑).
deliquescence (*n*) *see* **salt hydrate equilibria** (↑).
hygroscopic (*adj*) of a substance, having a tendency to
absorb water from the atmosphere and became damp,
but not form a solution.

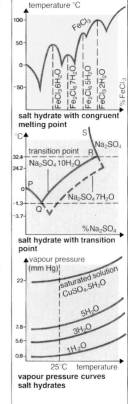

salt hydrate with congruent
melting point

salt hydrate with transition
point

vapour pressure curves
salt hydrates

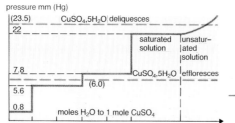

deliquescence and
efflorescence

_ _ _ atmospheric water
vapour pressure

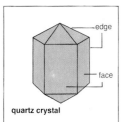

quartz crystal

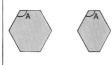

crystal habit

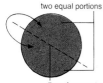

two equal portions

line of symmetry

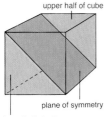

upper half of cube

plane of symmetry

lower half of cube

symmetry

crystal (*n*) a solid substance with a regular shape defined by naturally-formed plane faces. A particular substance always forms crystals of the same crystalline shape, and the crystals have a sharp melting point characteristic of the substance. **crystallization** (*n*), **crystallize** (*v*), **crystalline** (*adj*), **crystalloid** (*n*).

crystal structure a crystal has a regular external structure, its crystalline shape, and an internal structure. The internal structure is the regular arrangement in space of ions, atoms, or molecules of the crystalline substance.

crystal face a plane surface of a crystal, which intersects with other crystal faces at an angle. The angles between crystal faces are always constant; these are the interfacial angles (p.164).

crystal habit the external shape of a crystal. The shape of a crystal may vary owing to differing development of different faces, but the interfacial angle (p.164) between corresponding crystal faces (↑) remains constant.

crystal lattice the regular, geometrical arrangement in space of the ions, atoms or molecules, or their positions, in a crystal structure (↑). This is the internal structure of a crystal, it determines the crystal habit (↑). The crystal lattice can be ionic, metallic, molecular or giant molecular. It extends continuously in three dimensions to the boundary of the crystal, defined by the crystal faces (↑).

space lattice the lattice of a crystal (↑).

symmetry (*n*) the state of having a regular form such that a line or a plane divides a solid body into similar halves, e.g. a circle exhibits symmetry about any line drawn through its centre, a cube exhibits symmetry about any plane drawn through opposite edges. **symmetrical** (*adj*).

axis of symmetry a line drawn through a structure which divides the structure into two symmetrical halves.

n-fold axis of symmetry when a structure is rotated about an axis of symmetry, the same arrangement of the structure (called a congruent position) recurs *n* times in one complete rotation. If *n* = 2, the crystal must be rotated through 180°, and the axis is called a **diad** axis. If *n* = 3, the axis is a **triad**; if *n* = 4, the axis is a **tetrad**, if *n* = 6, the axis is a **hexad**. $n \neq 5$, $n \ngtr 6$. If *n* = 1, an **identity** axis is formed. The axes are also called two-fold, three-fold, etc.

plane of symmetry a plane drawn through a solid structure dividing it into two symmetrical halves.

crystallographic plane of symmetry in a crystal, a plane of symmetry that divides the crystal into two equal portions which are mirror images of each other.

point of symmetry any line drawn through a point of symmetry intersects the boundary, or surface, of a structure at equal distances in both directions from that point, e.g. the centre of a circle and the centre of a sphere are both points of symmetry.

centre of symmetry a point of symmetry (↑) in a crystal.

elements of symmetry the various possible types of symmetry, i.e. axes, planes and point, that a crystal exhibits, e.g. a cubic crystal has 13 axes, 9 planes and 1 centre of symmetry, making 23 elements of symmetry.

crystal symmetry a crystal can have one or more axes of symmetry (p.163) and one or more planes of symmetry (p.164), but it can have only one centre of symmetry (↑).

crystallography study of crystals. **crystallographic** (*adj*).

interfacial angle angle between two faces of a crystal.

law of constancy of angle in all crystals of the same substance the angles between corresponding faces have a constant value.

goniometer (*n*) an instrument that measures interfacial angles (↑). The simplest instrument is the contact goniometer. A reflecting goniometer makes accurate measurements on small crystals. **goniometry** (*n*).

unit cell a crystal is formed by the regular stacking of small identical units; these are unit cells, and they completely fill the space occupied by the crystal. Only seven types of unit cell are possible, and these seven cells define the seven crystal systems (↓). Each cell is defined by the length of three sides a, b and c, and three angles, α between b and c, β between a and c and γ between a and b.

crystal system crystals are grouped into seven major divisions, according to the type of unit cell (↑) in the crystal structure. The seven divisions can also be categorized from their basic axes of symmetry (p.165). The seven systems are: cubic, tetragonal, orthorhombic, rhombohedral, hexagonal, monoclinic and triclinic.

cubic system the unit cell (↑) of this system has three equal sides, and all angles between the sides are 90°. The system has 4 three-fold axes of symmetry (p.165). Examples of this system include crystals of sodium chloride, zinc blende, fluorite and diamond.

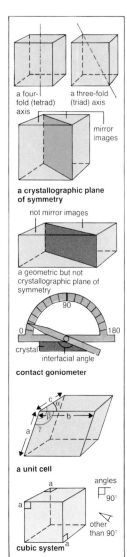

a four-fold (tetrad) axis

a three-fold (triad) axis

mirror images

a crystallographic plane of symmetry

not mirror images

a geometric but not crystallographic plane of symmetry

crystal

interfacial angle

contact goniometer

a unit cell

angles

90°

other than 90°

cubic system

tetragonal system

orthorhombic system

rhombohedral system

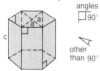

hexagonal system

monoclinic system

triclinic system

tetragonal system the unit cell (↑) of this system has two equal sides, with the third side of a different length. All angles between the sides are 90°. The system has at least one four-fold axis of symmetry (p.165). Examples of this system include crystals of tin (IV) oxide and titanium (IV) oxide.

orthorhombic system the unit cell (↑) of this system has three unequal sides, but all angles between the sides are 90°. The system has 3 two-fold axes of symmetry (p.165). Examples of this system include crystals of barium sulphate and rhombic sulphur.

rhombohedral system the unit cell (↑) of this system has three equal sides and three equal angles between the sides, but the angles are not 90°. The system has at least one three-fold axis of symmetry (p.165). Examples of this system include crystals of calcite and sodium nitrate.

trigonal system same as **rhombohedral system** (↑).

hexagonal system the unit cell (↑) of this system has two equal sides with a third side of different length. It has two equal angles of 90°; the third angle, between the two equal sides is 120°. The system has one six-fold axis of symmetry (p.165). Examples of this system include crystals of mercury (II) sulphide, and some metals, such as zinc and magnesium.

monoclinic system the unit cell (↑) of this system has three unequal sides. It has two equal angles of 90°; the third angle between the sides *a* and *c* is an angle other than 90°. This system has one two-fold axis of symmetry (p.165). Examples of this system include crystals of hydrated calcium sulphate and monoclinic sulphur.

triclinic system the unit cell (↑) of this system has three unequal sides and three unequal angles with none of them 90°. It has no axes of symmetry. Examples of this system include crystals of copper (II) sulphate and potassium dichromate (VI).

| crystal system | unit cell | | symmetry |
	sides	angles	
cubic	$a = b = c$	$\alpha = \beta = \gamma = 90°$	4 three-fold axes
rhombohedral	$a = b = c$	$\alpha = \beta = \gamma \neq 90°$	1 three-fold axis
tetragonal	$a = b \neq c$	$\alpha = \beta = \gamma = 90°$	1 four-fold axis
hexagonal	$a = b \neq c$	$\alpha = \beta = 90° \gamma = 120°$	1 six-fold axis
orthorhombic	$a \neq b \neq c$	$\alpha = \beta = \gamma = 90°$	3 two-fold axes
monoclinic	$a \neq b \neq c$	$\alpha = \gamma = 90° \beta \neq 90°$	1 two-fold axis
triclinic	$a \neq b \neq c$	$\alpha \neq \beta \neq \gamma \neq 90°$	no axes

face-centred cell a unit cell (p.166) with lattice points at
the cell corners, and lattice points at the centres of the
six cell faces. Only cubic and orthorhombic systems
have crystals formed with face-centred cells.

body-centred cell a unit cell (p.166) with lattice points at
the cell corners, and lattice points at the centre of each
cell. Cubic, tetragonal, orthorhombic and monoclinic
crystal systems have structures based on body-centred
cells.

close-packed structure in a crystal structure with non-
directive forms of bonding, the atoms or ions are in a
lattice with the most effective method of packing the
maximum number of particles in the smallest volume.
This forms a close-packed structure. The atoms or ions
are represented by spheres, and a single layer of
spheres is shown in the diagram, with a close-packed
structure. A second layer is added, with each sphere
resting in the space formed by three adjacent spheres.
A third layer can be added in two different ways forming
two different close-packed structures, the hexagonal
and the cubic close-packed structures (↓).

hexagonal close-packed structure in the diagram of
two close-packed layers, a third layer can be added at
point X, such that a sphere in the third layer is vertically
above the sphere in the first layer. This forms a
succession of layers which can be denoted by
ABABAB, i.e. the alternate layers occupy similar
positions. In any layer, a sphere is in contact with six
other spheres whose centres form a regular hexagon
coplanar with the central sphere. In addition, a sphere
is in contact with three spheres in the layer beneath it
and three in the layer above it. Hence any one sphere is
in contact with 12 other spheres. This is a hexagonal
close-packed structure.

cubic close-packed structure in the diagram of two
close-packed layers, a third layer can be added at
point Y, and such a sphere is not vertically above any
sphere in the two lower layers. This forms a succession
of layers which can be denoted by ABCABC, i.e. each
third layer occupies similar positions. As in the
hexagonal close-packed structure (↑) each sphere is in
contact with 12 other spheres. This is a cubic close-
packed structure.

body-centred cubic structure this is not a close-
packed structure, but some substances with close-
packed structures are polymorphic and can also

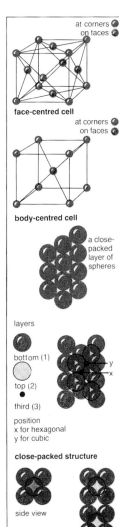

at corners ●
on faces ●

face-centred cell

at corners ●
on faces ●

body-centred cell

a close-
packed
layer of
spheres

layers

bottom (1)
top (2)
third (3)

y
x

position
x for hexagonal
y for cubic

close-packed structure

side view

**body-centred cubic
structure** top view

crystallize with this structure. A close-packed layer has 4 spheres in contact with one sphere in the next layer. Any one sphere is in contact with 8 other spheres.

coordination number the number of equidistant nearest neighbours of an atom or ion in a crystal lattice, or crystal structure. e.g. the coordination number of a hexagonal close-packed structure (↑) is 12. In ionic crystals, the coordination number is the number of nearest neighbours of opposite charge in contact with a given ion. The coordination number depends on the relative size of the ions; for ions of equal size, a coordination number of 12 should be possible, but ions are not found in this close-packed structure. The limiting value of the relative radii of ions for each type of coordination is shown in the table. Predictions from this table are not always correct, as the stability of a crystal lattice depends on energy considerations; the external conditions. e.g. temperature. can also alter crystal structure, as in enantiotropy (p.162). In crystals of non-metallic elements, the coordination number is usually equal to the normal valency of the element. In covalent crystals, the coordination number depends on the number of electrons available to form covalent bonds.

r_+ = radius of cation r_- = radius of anion

coordination number	coordination type	limiting value r_+/r_-
4	tetrahedral	0·225
6	octahedral	0·414
8	cubic	0·732

coordination number

interstice

interstice

tetrahedral coordination

octahedral coordination

interstice (n) a space between two or more solid structures. **interstitial** (adj).

interstitial site the small space between atoms or ions, i.e. the interstices (↑) between layers of atoms or ions in a crystal structure.

tetrahedral coordination an ion or atom in a crystal lattice, with 4 ions or atoms around the central ion or atom, arranged so that the centres form a regular tetrahedron. An ion in a tetrahedral site exhibits tetrahedral coordination.

octahedral coordination an ion or atom in a crystal lattice, with 6 ions or atoms around the central ion or atom, arranged so that the centres form a regular octahedron. An ion in an octahedral site exhibits octahedral coordination.

cubic coordination an ion or atom in a crystal lattice.
with 8 ions or atoms around the central ion or atom.
arranged so that the centres form a cube. An ion in a
body-centred cube exhibits cubic coordination.

coordination (n) the type of geometric arrangement of
ions in a crystal lattice. The interstitial sites (p.169) are
not necessarily all occupied. The coordination of cation
and anion are both expressed. In cadmium iodide
crystals a hexagonal close-packed structure of iodide
ions has half the octahedral sites occupied by cadmium
ions, alternate layers having fully occupied sites. The
coordination is 6:3, i.e. every cadmium ion has 6 iodide
ions around it. while every iodide ion has 3 cadmium
ions around it.

sodium chloride structure the unit cell is a primitive
cubic cell with alternate sodium and chloride ions at
each corner. Each sodium ion has 6 chloride ions as
neighbours. and each chloride ion has 6 sodium ions as
neighbours. The coordination is thus 6:6. A crystal can
be regarded as either: (a) two interpenetrating primitive
cubic lattices, one of sodium ions and one of chloride
ions. or (b) a cubic close-packed geometry of chloride
ions with sodium ions filling all the octahedral sites. The
structure is a close-packed geometry and not a close-
packed structure. as the sodium ions are too big to fit in
the interstitial sites (p.169) of a chloride ion lattice. so
the chloride ions are not in contact with each other. The
best description of the crystal structure is a cubic close-
packed type structure of chloride ions with sodium ions
occupying all the octahedral sites.

caesium chloride structure a body-centred cubic
close-packed type of structure of chloride ions with
caesium ions at the centre of the cube. It is not a true
body-centred close-packed structure since the body-
centre is not occupied by the same kind of atom as the
corners of the cube. and the ratio of the radii of caesium
ions to chloride ions is too great for a close-packed
structure (p.168). Each caesium ion is surrounded by 8
chloride ions and each chloride ion by 8 sodium ions.
so the coordination is 8:8. The structure is a cubic
close-packed type with cubic coordination.

zinc blende structure a cubic close-packed type of
structure of sulphide ions with zinc ions occupying half
the tetrahedral sites. The zinc ions have tetrahedral
coordination (p.169): the sulphide ions are also
surrounded tetrahedrally by zinc ions. The coordination

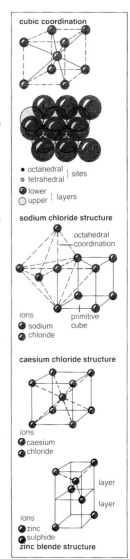

cubic coordination

● octahedral ⎱ sites
○ tetrahedral ⎰
◉ lower ⎱ layers
○ upper ⎰

sodium chloride structure

octahedral coordination

ions
◉ sodium
◉ chloride
primitive cube

caesium chloride structure

ions
◉ caesium
◉ chloride

layer
layer

ions
◉ zinc
◉ sulphide
zinc blende structure

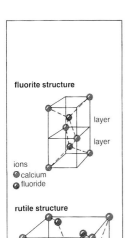

fluorite structure

layer

layer

ions
⊘ calcium
⊘ fluoride

rutile structure

⊘ titanium
⊘ oxide

**common ionic
crystal structures**

is 4:4. The bonds linking the ions have some covalent character. When arranged as a body-centred cubic cell, the structure has only half the corners occupied by sulphide ions, while alternate cells are occupied by zinc ions.

fluorite structure a cubic close-packed type of structure of calcium ions with fluoride atoms occupying all the tetrahedral sites. Each calcium ion has 8 fluoride ions arranged around it at the corners of a cube, and each fluoride ion has 4 calcium ions arranged tetrahedrally around it. The coordination is 8:4. Each cubic cell has only half the corners occupied by calcium ions, while each cube has a fluoride ion in a body-centred position.

rutile structure this can be described as a distorted body-centred cubic type of structure of titanium ions with oxide ions in a trigonal coplanar arrangement between the titanium ions. Each titanium ion has octahedral coordination, and each oxide ion a coordination number of 3. The coordination is thus 6:3. The structure is a distorted cube as one of the axes is shorter than the other two by about one-third.

iron pyrites structure iron pyrites (FeS_2) forms crystals with the same structure as that of sodium chloride (↑). The atoms of iron replace sodium, and two atoms of sulphur replace the chloride ions.

metal crystals most metals of the *s* and *d* blocks of the periodic table (p.50) have their ions arranged in one of the close-packed structures (p.168); cubic, hexagonal or body-centred cubic. Many metals are polymorphic and crystallize in more than one type of structure, usually at different temperatures. The crystal lattice consists of positive ions held in position by delocalized valency-shell electrons.

close-packed geometry		sites filled	coordination
cubic	hexagonal		
sodium chloride	nickel arsenide	all octahedral	6:6
zinc blende	wurtzite	half tetrahedral	4:4
fluorite	—	all tetrahedral	8:4
caesium chloride		cubic	8:8
cadmium chloride	cadmium iodide	half octahedral	6:3
chromium chloride		⅓ octahedral	6:2

interstitial compound a carbide, nitride, boride or hydride of a transitional metal. The crystals of an interstitial compound have the close-packed structure of the metal, with the non-metal atoms located in interstitial sites (p.169). Interstitial compounds have high melting points, are extremely hard due to dislocations formed by the interstitial element, exhibit the general physical properties of the metal, and generally are similar to alloys. Steel is an example of an interstitial compound, with carbon as the interstitial element.

molecular crystal a crystal structure in which atoms are joined by mainly covalent bonds to form individual molecules, and the molecules are held together by weak van der Waals', or hydrogen, bonds. Molecular crystals are soft with low melting points; they are non-electrolytes or insulators, e.g. iodine crystals, crystalline organic compounds.

giant molecular crystals crystals with a giant structure formed by covalent bonds. The three-dimensional lattice is determined by the availability of electrons to form spatially directed electron-pair bonds, e.g. carborundum, diamond, Al_2O_3. Giant molecular crystals are very hard with very high melting points; they are non-electrolytes or insulators, and generally insoluble in most solvents.

covalent crystals an alternative name for **giant molecular crystals** (↑).

ionic crystals crystals with a lattice of anions and cations, joined by non-directional ionic bonds. Ionic crystals are hard with high melting points; they are insulators when solid, but electrolytes when molten or in solution, e.g. NaCl, $CaCl_2$. In some ionic crystals, e.g. CdI_2, there may be a degree of covalency in the bonds, and the melting point of such crystals is lower than that of crystals with more ionic character.

cleavage (n) the property of a crystal that allows it to be split cleanly along certain planes, e.g. a cubic crystal can be split cleanly, i.e. cleaved, along planes parallel to a cube face. **cleave** (v).

cleavage plane a plane in a crystal along which the crystal can be cleaved (↑). The cleavage plane follows a direction in the crystal lattice in which the smallest number of ions or atoms is situated, i.e. a direction in which the bonding is weakest. Cleavage can take place along any one of a set of parallel planes. A crystal tends to fracture along its cleavage planes.

interstitial compound

metal lattice

metal positive ions

interstitial non-metal atom

giant molecular crystal structure

ions
🔘 aluminium
⬤ oxide

cleavage plane

cubic crystal

cleavage plane

substance	melting point	bond type	characteristics		structure
sodium chloride	808°C	ionic	hard crystals	electrolyte	giant
caesium chloride	646°C	ionic	hard crystals	electrolyte	giant
cadmium chloride	568°C	ionic layer some covalency	flaky crystals	electrolyte	layer lattice
palladium chloride	500°C decomposed	covalent coordinate	needle-shaped crystals	electrolyte	molecular chain
aluminium oxide	2050°C	covalent	very hard crystals	insulator	giant
iodine	114°C	covalent	soft crystals	non-electrolyte	molecular
naphthalene	80°C	covalent	soft crystals	non-electrolyte	molecular
iron	1539°C	metallic	malleable crystals	conductor	close-packed
tin	232°C	metallic	malleable crystals	conductor	close-packed

crystal types

isotropic (*adj*) describes crystals which are identical in all directions. Cubic crystals (p.166) and amorphous substances, such as glasses and plastics, are isotropic.

Born-Haber cycle the lattice energy of a crystalline substance can be calculated from the Born-Haber cycle *see diagram*, which shows the calculation for sodium chloride. The process of forming the lattice can be broken down into stages, and the stages, marked S, D, I and F, can be measured experimentally. If E is known, then ΔH can be calculated. If ΔH is calculated from the Born-Mayer equation, then E can be calculated. ΔU is calculated from $\Delta H = \Delta U - 2RT$, where ΔU is the lattice energy.

Born-Haber cycle for sodium chloride

S = heat of sublimation
D = heat of decomposition
I = heat of ionization
E = electron affinity
F = heat of formation
ΔH = lattice energy

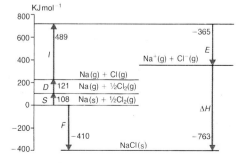

disperse systems systems in which fine particles of a solid, very small droplets of a liquid or small bubbles of a gas, the disperse phase (↓), are dispersed throughout another liquid or gas, the dispersion medium (↓).

dispersion medium the gas or liquid in which a solid, liquid or gaseous disperse phase (↓) is uniformly distributed throughout its volume. The dispersion medium retains a continuous phase structure, whilst that of the dispersion phase is discontinuous.

disperse phase any solid, liquid or gas in a colloidal state (↓) that is dispersed throughout a dispersion medium (↑).

dispersed (*adj*) refers to any substance distributed in the form of fine colloidal particles or droplets throughout a disperse phase (↑).

colloidal state a system in which fine particles of a solid, or droplets of a liquid, with diameters less than 100 nm, are uniformly dispersed (↑) throughout a liquid or solid.

colloidal system any system consisting of a colloid (↓) distributed throughout a dispersion medium (↑), e.g. milk, foam (↓), aerosols.

colloid (*n*) the non-continuous disperse phase (↑) uniformly distributed throughout a dispersion medium (↑). Particles of a colloid are small enough to pass through conventional filters, but not through semipermeable membranes, *see* **Tyndall effect** (p.176). **colloidal** (*adj*).

phase ratio the ratio of the mass of the disperse phase to that of its dispersion medium (↑) expressed as a percentage.

intrinsic colloids refers to colloids formed from solid materials which form colloidal dispersions when warmed or brought into contact with a suitable dispersion medium (↑), e.g. rubber in benzene, or long chain (p.33) macromolecules with polar groups, such as soaps.

extrinsic colloids refers to insoluble solid materials of low relative molecular mass (p.29) which have to be specially treated to form sols (p.176), e.g. gold sols. These metal sols are lyophobic (p.176) in character.

foam (*n*) a disperse system (↑) consisting of small bubbles of gas dispersed (↑) throughout a liquid. Each bubble of gas is enclosed within a thin film of the liquid. Some foams can be allowed to solidify leaving the gas trapped inside a solid matrix. Foams are made by passing gases under pressure through liquids, or by forming the gas by chemical reaction in solution. **foam** (*v*).

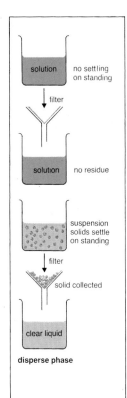

disperse phase

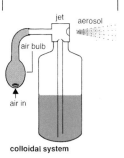

colloidal system

froth (*n*) an unstable type of foam (↑) that consists of large bubbles of gas and tends to break down easily. Air forms a froth with detergent (↓) film on water, and carbon dioxide forms a froth when carbonated drinks are poured from bottles. **froth** (*v*).

smoke (*n*) a dispersion of fine solid particles in a gas, especially of combustion products in air, e.g. the residual solids released from factory chimneys which are too small and light to settle out. **smoke** (*v*).

emulsion (*n*) a disperse system (↑) consisting of two immiscible liquid phases in which one liquid in the form of fine droplets is dispersed throughout the other liquid. Many emulsions will separate into the two individual phases unless the emulsion is stabilized by use of an emulsifying agent (↓). Milk is an emulsion of fat droplets in whey and separates on standing.

emulsifying agent any substance which when added to an emulsion (↑) will stabilize it and help to prevent the components of the emulsion from separating out. Substances which reduce the surface tension between the liquids are used as emulsifying agents, e.g. soaps (↓) and detergents (↓).

wetting agent a surface active agent used specifically for reducing the surface tension of water so that it will wet and spread over the surface of a solid more readily.

detergent (*n*) any of a group of surfactants similar to soaps (↓), consisting of a long carbon chain (p.33) joined to a sulphonic acid or phosphate group. The carbon chain is oil soluble, hydrophobic (p.176), and the inorganic portion is water soluble, hydrophilic (p.176). The molecules of the detergent distribute themselves at the interface between the oil layer and the water layer, thus reducing the surface tension between the two. They are used as cleaning agents, but unlike soaps (↓) are not affected by hard water.

fat droplets in water

emulsion

fat soluble
hydrophobic end

$$CH_3(CH_2)_{16} \, CO\overset{-}{O} \, \overset{+}{Na}$$

water soluble
hydrophilic end

a soap—sodium stearate

a detergent—sodium lauryl sulphonate

$$CH_3(CH_2)_{11} \, O.SO_2.\overset{-}{O} \, \overset{+}{Na}$$

fat soluble hydrophobic end

water soluble
hydrophilic end

soap (*n*) sodium and potassium salts of long chain (p.33) carboxylic acids, e.g. potassium laurate, sodium oleate, sodium stearate. They are surfactants, reduce the surface tension of water and are used as cleaning agents. Unlike detergents (↑) they interact with calcium and magnesium salts dissolved in hard water and form a scum on the surface of the water. **soapy** (*adj*).

association colloid a disperse system formed under the effect of a soap (p.175) or detergent (p.175) in which a very fine mixed colloid consisting of aggregates of small numbers of molecules is created.

sol (*n*) refers to any colloidal system (p.174) in which the dispersion medium (p.174) is a liquid. *See also* **lyophilic sol** (↓) *and* **lyophobic sol** (↓).

lyophilic sol a sol (↑) consisting of a disperse phase (p.174) which has an affinity for the continuous phase. This means that the colloid is readily formed, e.g. starch in water, albumin in water.

lyophobic sol refers to a sol (↑) which is solvent-repelling such that the disperse phase has little or no attraction for the dispersion medium (p.174). As a result the colloid is easily broken, e.g. gold in water.

hydrophilic (*adj*) refers to a lyophilic sol (↑) in an aqueous (water-based) dispersion medium.

hydrophobic (*adj*) refers to a lyophobic sol (↑) in an aqueous (water-based) dispersion medium.

hydrosol (*n*) a sol in which the dispersion medium is water, e.g. a sulphur hydrosol in which the disperse phase (p.174) is fine sulphur particles and the dispersion medium is water.

aquasol (*n*) another name for a **hydrosol** (↑).

gel (*n*) an intermediate stage in the conversion of a sol (↑) to a suspension prior to its coagulation. It is a semi-solid mass formed due to the creation of intertwining threads by the colloid thus enclosing the dispersion medium (p.174). The resulting substance is a mobile, readily deformed jelly. **gel** (*v*), **gelatin** (*n*).

coagulation (*n*) the formation of large, non-dispersed particles from a colloid. Some colloids are coagulated by the addition of chemicals, by vibration, electric current, or heat, e.g. alum is used to coagulate colloidal substances in sewage in order to clarify the water.

isoelectric point the pH value (p.132) at which a sol (↑) will not move towards either the anode or the cathode when in an electric field, because at that pH it carries no electric charge, e.g. the isoelectric point for casein in milk is a pH of 4.1 to 4.7.

Tyndall effect the observation that a ray of light is scattered by small particles in the path of the beam. This occurs with colloids in a disperse medium (p.174), and with dust particles in gases. The beam itself becomes visible as a result of striking and being reflected from the surface of the particles.

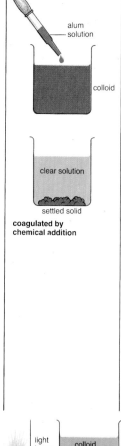

coagulated by chemical addition

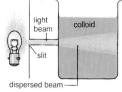

Tyndall effect

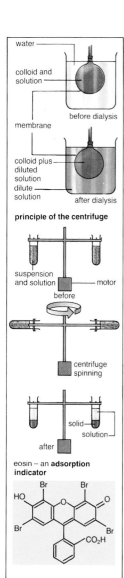

water

colloid and solution

before dialysis

membrane

colloid plus diluted solution

dilute solution

after dialysis

principle of the centrifuge

suspension and solution — motor

before

centrifuge spinning

solid — solution

after

eosin – an **adsorption indicator**

Tyndall beam refers to a beam of light made visible due to the Tyndall effect (↑) on passage through a colloid solution. A parallel beam of light passing through the colloid is progressively spread out as it strikes particles and becomes cone-shaped.

stabilize (*v*) to make stable, improve stability or prevent change. Sols (↑) are stabilized by the addition of small quantities of other chemicals, **stabilizer** (*n*).

dialysis (*n*) in general, refers to the separation of mixtures as a result of the selective diffusion of their components through semipermeable membranes. It is a procedure especially used to separate a sol (↑) and a true solution from each other as the ions and molecules in the solutions will diffuse much more easily and rapidly through a membrane.

crystalloid (*n*) any substance which is able to form crystals and produces a true solution in a solvent from which it can be recovered in a crystalline form. A crystalloid can be separated from a sol by dialysis (↑). **crystal** (*n*), **crystallization** (*adj*).

electrophoresis (*n*) describes the movement of charged particles, especially colloids, through a stationary gel or liquid due to the effect of an applied electric field. It is used particularly for separating and identifying the components in mixtures of amino acids (p.41).

cataphoresis (*n*) another name for **electrophoresis** (↑).

centrifugation (*n*) a method of separating a solid precipitate or suspension from a liquid by increasing its sedimentation rate. It is achieved by the centrifugal force applied to the solid when the solid/liquid mixture is turned at high velocity in a container at the end of a rotating horizontal arm. **centrifuge** (*n*), **centrifugal** (*adj*).

electrostatic precipitation a process for the removal of small particles of waste solids (and sometimes liquids) suspended in air or waste gases, by attracting them to a series of charged, thin wires from which the particles are periodically shaken and then attracted into an electrically charged hopper. The process is used to obtain a very high degree of purification of gaseous effluents before they are released into the atmosphere.

adsorption indicator refers to organic dyes, e.g. eosin and fluorescein, which are adsorbed on to the surface of precipitates produced near the equivalence points of some titrimetric (p.137) analytical procedures. The change of colour due to this adsorption serves to indicate the end-point of the titration.

macromolecular system (1) a crystalline form of an element or compound which appears to be continuous, and in which individual molecular units cannot be identified, e.g. diamond may be regarded as a single, large carbon molecule. (2) a molecule consisting of a large number of regular, repeating units, usually in large chains or networks, e.g. starch, cellulose, proteins.

surface phenomena a general term referring to phenomena, occurring at surfaces and interfaces, due to the properties of solids, liquids and gases. Surfaces and interfaces do not possess the same properties as the bulk of the material does, as the interatomic and intermolecular forces at the surface are not uniformly distributed. Properties involving surface characteristics include surface tension (p.154), friction, adsorption (↓) and electric double-layers.

surface film (1) a thin layer, often one molecule thick, of an insoluble or partially soluble liquid on the surface of another, e.g. oil on water, a film of detergent. (2) a thin surface layer formed on a sheet of metal due to the action of atmospheric gases such as oxygen or carbon dioxide, e.g. silver tarnishes due to the formation of a surface film of silver sulphide.

adsorption (*n*) a surface phenomenon (↑) in which a thin layer or film of a gas, vapour or liquid is held on the surface of a solid due to a combination of physical and weak chemical forces. The film of the adsorbed species may only be one or two molecules thick. This process differs greatly from that of **absorption** in which strong chemical bonds (p.69) are formed.

physical adsorption an adsorption (↑) process, involving only physical forces, that is readily reversible. The adsorption is normally due to van der Waals' forces and decreases with increased temperature, e.g. ammonia or methane adsorbed on the surface of charcoal in which low heats of adsorption of about 5-10 kcal per mole are involved.

van der Waals' adsorption *see* **physical adsorption** (↑).

chemisorption (*n*) an adsorption process involving chemical forces between the adsorbent (↓) and the gas or vapour being adsorbed. Chemisorption involves the formation of chemical bonds so that the heats of adsorption involved are greater than those in physical adsorption (↑), being 10–100 kcal per mole. The extent of chemisorption increases with increased temperatures, e.g. oxygen on charcoal.

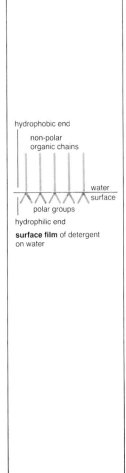

hydrophobic end

non-polar organic chains

water

surface

polar groups

hydrophilic end

surface film of detergent on water

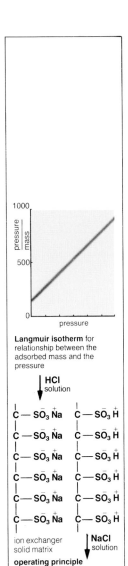

Langmuir isotherm for relationship between the adsorbed mass and the pressure

operating principle of a cation exchanger

activated adsorption *see* **chemisorption** (↑).

adsorbent (*n*) any substance which will adsorb gases and vapours by relatively weak forces on its surface; especially porous materials, possessing large surface areas, e.g. charcoal, silica gel, platinum.

heat of adsorption the heat evolved when a gas or vapour is adsorbed on an adsorbent (↑). The adsorption process is associated with a decrease in heat content; it can be defined according to the adsorption isotherm.

monolayer (*n*) refers to a uniform surface film (↑) with a thickness of a single molecule, it is also called a **monomolecular layer.**

Langmuir adsorption isotherm a mathematical relationship between the quantity of gas or vapour adsorbed and the equilibrium pressure (measured at constant temperature), based upon the concept that the adsorbed gas has not formed more than a monolayer (↑) on the adsorbent (↑) surface. The equation for the relationship is such that a graphical plot of the pressure divided by the adsorbed mass against the pressure gives a straight line.

ion exchange (1) any process in which one ion is exchanged for another, e.g. in the addition of silver nitrate to a sodium chloride solution, the silver is exchanged for the sodium and silver chloride is precipitated. (2) any process involving a solid polymeric matrix in which ions of one species held on the matrix are replaced by ions of another species in a solution passing over the surface of the solid, which is called an ion exchange resin.

cation exchange the exchange of one cation (p.126) for another using a cation exchange resin in which one species of cation, held on the resin, is displaced and replaced by a second species of cation in a solution passed through the resin. The ion exchange resin consists either of fine granules packed in a glass column, or may be in a sheet or membrane (p.118) form, e.g. calcium ions in hard water can be replaced by sodium ions by passing the water down a cation exchange column containing sodium ions in the matrix.

acid exchange refers to the use of a cation exchanger possessing protons (p.44) in its matrix which can be exchanged for other cations in solutions passed down the exchanger, such that the protons are released from the column giving acidic solutions.

anion exchange the exchange of one anion (p.126) for another using an anion exchange resin in which one species of anion held on the ion exchange resin is displaced and replaced by a second species of anion in a solution passed through the resin, which may be packed in a glass column or in the form of a sheet or membrane (p.118), e.g. chloride ions in brine can be replaced by sulphate ions by passing the brine down an anion exchange column containing sulphate ions in the matrix.

ion exchanger
solid matrix

$C — NR_3 Cl$

$C — NR_3 Cl$

$C — NR_3 Cl$

$C — NR_3 Cl$

Na OH
solution

**operating principle of
an anion exchanger**

$C — NR_3 OH$

$C — NR_3 OH$

$C — NR_3 OH$

$C — NR_3 OH$

NaCl
solution

base exchange refers to the use of an anion exchanger possessing hydroxyl ions in its matrix which can be exchanged for other anions (p.126) in solutions passed down the exchanger, such that hydroxyl ions are released from the column giving basic solutions.

deionization (*n*) the total removal of cations (p.126) and anions (p.126) from water, and their replacement by hydrogen ions H^+ and hydroxyl ions OH^- respectively, such that the water is purified of foreign ions. Deionization is achieved by subjecting any water contaminated by metal salts in solution to an acid exchange (p.179) process, simultaneously with a base exchange (↑) process, through a column in which the two necessary ion exchange (p.179) resins are mixed together. **deionizer** (*n*).

demineralization (*n*) another term for **deionization** (↑).

ion exchange membrane a thin, flexible sheet of material about 0.05 cm thick formed from either an anion exchange (↑) resin or a cation exchange (p.179) resin. These membranes are used for dialysis (p.177) and for electro-osmosis.

selective adsorption refers to any adsorption process in which a surface adsorbs and retains one chemical species in preference to another. Adsorption chromatography (p.23) is dependent upon selective and preferential adsorption enabling one chemical species to be separated from others.

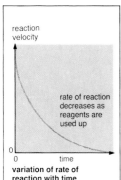

reaction
velocity

rate of reaction
decreases as
reagents are
used up

0

0   time

**variation of rate of
reaction with time**

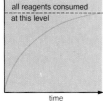

amount of reagent
used (moles/dm^3)

all reagents consumed
at this level

time

**change in reagent
concentration with time**

reaction kinetics the study of the rates at which
chemical reactions proceed, and the influence of such
factors as temperature, pressure and concentrations of
reactants upon the rates of reaction.

rate of reaction a measure of the rate at which either
reactants are used in a reaction, or products are
formed. This rate depends upon a variety of conditions
including temperature, pressure, catalysis (p.187),
nature and concentration of reactants. The rate of
reaction decreases with time as the concentration of
reactants diminishes.

specific reaction rate *see* **rate constant** (↓)

velocity constant *see* **rate constant** (↓).

rate constant refers to a constant numerical value in a
rate equation for a particular chemical reaction, e.g. in a
simple reaction in which a single reactant is used up,
the rate of reaction dz/dx at any time is proportional to
$x - z$, where x is the initial concentration and z is the
quantity used up at that point. This can be expressed
as: $dz/dx = k_1(x - z)$. In this equation k_1 is the rate
constant for the particular reaction carried out under the
specified reaction conditions.

law of Guldberg and Waage *see* **law of mass action** (↓).

law of mass action states that the rate at which a
chemical reaction proceeds is proportional to the active
masses of the reacting substances. So that in a reaction
expressed as: $A + B \rightarrow Z$
                    reactants   product
the rate of the reaction between A and B is proportional
to the molar concentrations of the two substances, such
that: rate of reaction = $k[A][B]$.

active mass in the law of mass action (↑), active mass
refers to molar concentrations for substances in
solution, or partial pressures of gaseous reactants.

activity^3(*n*) a general assessment of an element's ability
and tendency to react with other elements and
compounds based upon the number of reactions it will
readily undergo, e.g. fluorine has a high activity as it
reacts vigorously with many other substances, but gold
has a very low activity.

activity coefficient a means of allowing for, and
correcting the deviation of a solution from ideal
behaviour, whereby the concentration of the solute is
multiplied by a factor known as the activity coefficient,
f_B, such that in equations normally using the molar
concentration [A] the term $f_B[A]$ is used instead.

partial pressure refers to a mixture of gases, the total pressure of which is equal to the sum of the pressures which each gas would exert if it occupied the entire volume by itself. Such that the total pressure P for a mixture of three gases A, B, and C is given by the sum of the three partial pressures: $P = p_A + p_B + p_C$.

fugacity (n) a thermodynamic quantity, symbol f, used to correct the non-ideal behaviour of a gas or vapour above a liquid. It is a measure of the tendency of a liquid or substance in solution to escape as a vapour.

fugacity coefficient the ratio of the fugacity (↑) of a gas to its pressure. The value of f/p is unity for ideal gases (p.147) and for gases at low pressures. The greater the deviation from ideal behaviour the larger the value of f/p.

chemical potential the free energy possessed by one mole of a substance.

Arrhenius equation an equation relating the variation of rate constants (p.181) with temperature. It may be expressed in several forms, one of which is:
$k = Ae^{-E/RT}$, where k is the rate constant, A is the Arrhenius factor, and E the activation energy of the system.

Arrhenius factor a constant used in the Arrhenius equation (↑) relating to the variation of rate constant (p.181) with temperature for a particular reaction. ·

frequency factor another name for the **Arrhenius factor**.

pre-exponential factor same as **Arrhenius factor**(↑).

order of reaction refers to the number of chemical species, i.e. atoms, molecules or ions, whose concentrations determine the kinetics and rate at which a reaction proceeds.

zero order reaction when the rate of a reaction does not vary with the concentration of a particular reactant, such a reaction is said to be of zero order with respect to that reactant, e.g. the decomposition of ammonia on the surface of metals.

first order reaction the rate of reaction proportional to the first power of the concentration of a single substance, e.g. the decomposition of nitrogen (V) oxide is a first order reaction as the rate of the reaction depends only upon the concentration of the nitrogen (V) oxide. Most first order reactions are decompositions of gaseous compounds.

$$N_2O_5 \longrightarrow N_2O_4 + \tfrac{1}{2}O_2$$

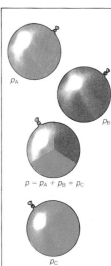

p_A

p_B

$p = p_A + p_B + p_C$

p_C

total pressure is the sum of the individual pressures

the first order decomposition of nitrogen pentoxide

pseudo-first order reaction refers to any reaction which apparently follows first order reaction (↑) equations, but which involves more than one chemical species. This occurs in reactions in which one of the species is present in great excess compared with the others, such that its concentration remains practically constant throughout the reaction.

second order reaction the rate of reaction is proportional to the product of the concentrations of two substances, or the square of the concentration of a single substance, e.g. the hydrolysis of an ester (p.40) depends upon the concentration of the ester and the hydroxyl ions:
$$CH_3COOC_2H_5 + OH^- \rightarrow CH_3COO^- + C_2H_5OH.$$

the second order hydrolysis of an ester

$$CH_3COOC_2H_5 + \overline{O}H \longrightarrow$$
$$CH_3CO\overline{O} + C_2H_5OH$$

third order reactions these are uncommon, but have a rate equation of the type: rate = $k[A]^3$ or rate = $k[A]^2[B]$ or rate = $k[A][B][C]$, e.g. $2NO + O_2 \rightarrow 2NO_2$.

$$2NO + Cl_2 \longrightarrow 2NO\,Cl$$

the third order reaction between nitric oxide and chlorine

overall order some chemical reactions proceed by a series of steps each of which will be a first order (↑), second order or third order reaction; the overall order of the reaction will depend upon the nature of these individual steps. The overall order of a reaction is the sum of the indices of all the reactants whose concentrations are found experimentally in the rate equation. This is usually, but not always, a whole number.

half-life (n) in terms of reaction kinetics (p.181), the half-life is the time required for one half of the reactants to have undergone decomposition under the reaction conditions. The half-life period for a reactant undergoing a first order reaction (↑) is a constant independent of the initial concentration.

quantity of
material remaining

quantity remaining is halved in each interval of time ($t_{1/2}$)

$1/2$
$1/4$
$1/8$
$1/16$

1   2   3   4   5
equal intervals of time

half-life

$$R\,Cl$$
↓ slow
$$\overset{+}{R} + \overset{-}{Cl}$$

$$\overset{+}{R} + \overset{-}{OH}$$
↓ fast
$$R\,OH$$

the slowest step determines the rate of the overall reaction

the rate determining step

rate determining step in any chemical reaction proceeding by more than one stage the overall rate of the reaction is determined by the slowest stage, this is known as the rate determining step, e.g. in the hydrolysis of alkyl halides (p.37) the rate determining step is that involving the ionization of the alkyl halide as that is the slower of the two steps in the reaction sequence.

molecularity (*n*) refers to the number of molecules reacting together according to the stoichiometric equation of the reaction. The molecularity is not necessarily identical with the order of the reaction.

unimolecular (*n*) a chemical reaction involving the decomposition of a single molecule in the stoichiometric equation for the reaction, e.g. the decomposition of malonic acid.

$$CH_2 \overset{\displaystyle COOH}{\underset{\displaystyle COOH}{\Big\langle}} \longrightarrow CH_3COOH + CO_2$$

the unimolecular decomposition of malonic acid

bimolecular (*n*) a chemical reaction involving the interaction of two molecules according to the stoichiometric equation for the reaction, e.g. the reactions between alkyl halides (p.37) and amines (p.38) to give quaternary ammonium halides.

$$RCl + R'NH_2 \longrightarrow \overset{\displaystyle R}{\underset{\displaystyle R'}{\Big\langle}} \overset{+}{N}H_2 \overset{-}{Cl}$$

the bimolecular formation of a quaternary ammonium halide

trimolecular (*n*) a chemical reaction involving the interaction of three molecules according to the stoichiometric equation for the reaction, e.g. the decomposition of potassium chlorate (I) into potassium chlorate (IV) and potassium chloride.

$$3KClO \longrightarrow KClO_3 + 2KCl$$

the trimolecular decomposition of potassium hypochlorite

activated complex an intermediate chemical structure of a high energy state initially formed by chemical reactants before it decomposes to the required products. It is represented in a chemical equation as a formula enclosed in square brackets, in the following form:

A + B   →   [A.B]   →   products
reactants   activated
              complex

activated state *see* **activated complex.** (↑).

transition complex an **activated complex** (↑).

$$N_2 + 3H_2$$

$$\updownarrow$$

$$2NH_3$$

the reversible reaction for the formation of ammonia

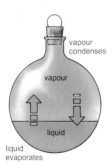

vapour condenses

vapour

liquid

liquid evaporates

dynamic equilbrium of chloroform liquid with its vapour

kinetic equation (1) a general equation for gases based upon kinetic theory (p.145) and concepts of mechanics, which has the form: $pV = \frac{1}{3}mn\bar{u}^2$, in which the pressure, p, and volume, V, of the gas are related to the molecular mass, m, the number of molecules, n, and the root mean square velocity of the molecules. This equation may be used to derive the gas laws (p.148). (2) refers to any equation used to determine the reaction kinetics (p.181) and rate constants (p.181) in a chemical reaction.

reversible reaction a chemical reaction which may proceed in either direction depending upon the applied conditions, e.g. the reaction between hydrogen and iodine to give hydrogen iodide is nearly complete at 450°C, but the hydrogen iodide gas decomposes back to the constituent elements at a lower temperature: $H_2 + I_2 \rightleftharpoons 2HI$

chemical equilibrium a situation in a reversible reaction (↑) when the rate of the forward reaction is equal to the rate of the reverse reaction such that products are being formed at the same rate as they are being decomposed.

dynamic equilibrium describes a chemical equilibrium (↑) which is maintained despite continual change between molecules, usually in a closed system, e.g. the equilibrium between nitrogen dioxide and its dimer dinitrogen tetroxide: $N_2O_4 \rightleftharpoons 2NO_2$.

equilibrium constant for any reaction, the ratio of the concentrations of the products multiplied together divided by the concentrations of the reactants is a constant at a specified temperature. In mathematical terms the equilibrium constant for the reaction: $mA + nB \rightleftharpoons yC + zD$ is given by:

$$K_c = \frac{[C]^y[D]^z}{[A]^m[B]^n}$$

The square brackets represent molar concentrations. K_c is the symbol for the equilibrium constant; K_p is the symbol for the equilibrium constant in reactions involving gaseous reactants and products in which partial pressures may be used in place of molar concentrations; K_f is the symbol for the equilibrium constant if activities are used in place of molar concentrations; K_f is then known as the thermodynamic equilibrium constant.

equilibrium mixture a mixture of reactants and products in a reversible reaction which have attained an equilibrium state.

homogeneous reaction refers to any reaction which takes place entirely in a single phase, either gaseous or in solution.

heterogeneous reaction refers to any reaction involving two or more separate phases, e.g. a hydrogenation reaction involves gaseous hydrogen shaken with an organic solution in the presence of a solid nickel catalyst ($\downarrow$).

dissociation pressure the partial pressure in a closed vessel exerted by a gas formed in a reversible reaction by the dissociation of a solid, e.g. the partial pressure of carbon dioxide formed from calcium carbonate:

$$CaCO_3 \rightarrow CaO + CO_2$$
solid   solid   gas

The dissociation pressure is independent of quantities of the two solids and varies only with temperature.

van't Hoff isotherm a mathematical expression relating the free energy of a system to the equilibrium constant (p.185) for the reaction and the concentration of reactants and products at any specified stage. The equation takes the form: $\Delta G = RT\ln K + RT\ln Q_d$, and at a state of chemical equilibrium (p.185): $K = Q_d$ so $\Delta G = 0$.

standard isotherm refers to the van't Hoff isotherm ($\uparrow$) calculated in terms of standard states, which are pressures of one atmosphere and solutions of one mole per litre. The van't Hoff equation then becomes:

$$\Delta G = RT\ln Q_c/K_c.$$

ideal standard isotherm the van't Hoff isotherm employed in connection with gases with which it is usual practice to employ the state of unit activity (p.181) for an ideal gas at one atmosphere. Under these conditions the equation for gaseous reactions becomes: $\Delta G = -RT\ln K_p$.

free energy change refers to the change in the energy in a chemical system available to do work. This is calculated in terms of the Gibbs free energy, which is a property of state relating enthalpy (p.112), entropy (p.113) and temperature. For processes at constant temperature the change in free energy is given by the equation: $\Delta G = \Delta H - T\Delta S$.

van't Hoff isochore another name for the **van't Hoff isotherm** ($\uparrow$).

Le Chatelier's principle refers to reversible reactions which can exist in equilibrium, and states that if there is any alteration in one of the factors maintaining the

Le Chatelier's principle illustrated by the formation of increasing yields of ethyl acetate (ethyl ethanoate) by increasing the proportion of ethanol

reactants	product
alcohol	ester
0·2 mole	0·2 mole
0·3	0·3
0·5	0·4
1·0	0·7
2·0	0·9
8·0	1·0

equilibrium of the system, e.g. temperature or pressure,
the equilibrium, will alter in order to minimize the effect
of the alteration as far as possible, e.g. in the formation
of nitrogen monoxide from nitrogen and oxygen heat is
required:

$$N_2 + O_2 \rightleftharpoons 2NO \qquad \Delta H = + 180 kJ \text{ per mole}$$

As a consequence raising the temperature helps to
move the equilibrium to the right and increases the yield
of nitrogen monoxide.

chain reaction any chemical reaction which proceeds
by a series of steps in which the molecule, radical or ion
(p.124) initiating the process is regenerated at the end
of a cycle of reactions such that the process can be re-
initiated, e.g. the reaction of chlorine with hydrogen
when exposed to light proceeds by a chain reaction in
the following steps:

$$(1) \, Cl_2 \rightarrow 2Cl\cdot$$
$$(2) \, Cl\cdot + H_2 \rightarrow HCl + H\cdot$$
$$(3) \, H\cdot + Cl_2 \rightarrow HCl \, Cl\cdot$$

The chlorine atom at the end of step (3) continues the
chain reaction by further reacting with a hydrogen
molecule. The process will only terminate when either
reagent is exhausted or inhibitors (↓) are present.

chain carrier atoms, molecules or radicals which take
part in and enable a chain reaction (↑) to continue.

inhibitor (*n*) a substance which removes the atoms and
radicals acting as chain carriers (↑), such that a chain
reaction (↑) cannot continue. **inhibit** (*v*), **inhibition** (*n*).

catalysis (*n*) the process of adding a catalyst (↓) to a
reaction to alter the rate at which the products are
formed. **catalyze** (*v*).

catalyst (*n*) any substance which, when added to a
reaction, alters the rate of the reaction but remains
chemically unchanged at the end of the process,
e.g. vanadium (v) oxide acts as a catalyst to increase
the rate of combination of sulphur dioxide and oxygen
to produce sulphur (VI) oxide. Most catalysts are
metals, such as nickel, platinum and palladium, in a
very finely divided form with a large surface area.
catalytic (*adj*).

**the catalytic hydrogenation
of propene**

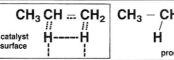

$$CH_3 - CH = CH_2$$
$$H - H$$
reactants

$$CH_3 \, CH = CH_2$$
catalyst $\quad$ H------H
surface

$$CH_3 - CH - CH_2$$
$$ H \qquad H$$
products

negative catalyst a substance which when added to a reaction slows down the rate (1) by directly interfering with the reaction mechanism; or (2) by reacting with a catalyst and destroying its action.

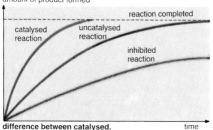

amount of product formed

reaction completed

catalysed reaction

uncatalysed reaction

inhibited reaction

difference between catalysed, uncatalysed and inhibited reactions — time

autocatalysis (*n*) refers to a chemical reaction in which one of the products itself serves as a catalyst (p.187). As a result the reaction is initially slow until some of the catalyst is formed and then increases as the catalyst concentration increases. Oxidation reactions involving potassium manganate (VII) are often autocatalytic due to the formation of Mn^{2+} ions. **autocatalytic** (*adj*).

promoter (*n*) a substance which is not itself a catalyst (p.187) but greatly increases the catalytic effect when mixed with the catalyst, e.g. aluminium (III) oxide added to iron greatly improves the rate of combination of nitrogen and hydrogen to form ammonia.

inhibitor (*n*) any substance which inhibits or retards the action of a catalyst (p.187), *see* **negative catalyst** (↑). **catalytic poisons** (↓).

accelerator[2] (*n*) refers generally to any substance which may be added to a mixture of reagents to increase the rate of the chemical reaction, especially to a catalyst.

activator (*n*) any chemical substance which assists in initiating a chemical reaction but does not act as a catalyst by acting throughout the reaction or remaining unchanged, e.g. a couple of crystals of iodine may be used to initiate the formation of a Grignard reagent, but the iodine is consumed in the process.

mixed catalysts the use of a mixture of two catalysts where this is more effective than the use of just a single catalyst, e.g. zinc oxide and chromium oxides are mixed together to catalyze the reduction of carbon monoxide to methanol.

catalytic poisons substances present in small
quantities or added to catalysts (p.187) to occupy the
active centres ($\downarrow$) and reduce, or totally inhibit, the
normal catalytic action. Sulphur compounds and
arsenical compounds are particularly strong poisons
for metal catalysts such as nickel and palladium.

retardation (n) the decreasing of the rate of a catalyzed
chemical reaction due to one of the reactants or one of
the products being so strongly adsorbed (p.178) on the
active centres ($\downarrow$) of the catalyst (p.187) that the
remaining reactants cannot reach the catalyst surface.

retarder (n) a substance added to a reaction in order to
slow down the rate of formation of products. This may
be in the form of a catalytic poison ($\uparrow$), or simply a
diluent effect as a result of, for example, adding an inert
gas to a gaseous reaction mixture.

$$\frac{1}{2}O_2 + NO$$

$$\downarrow \text{ catalyst}$$

$$NO_2$$

$$NO_2 + CO$$

$$\downarrow$$

$$CO_2 + NO$$

catalyst
recycled

**homogeneous gaseous
phase catalysis** of the
oxidation of carbon monoxide

active centre refers to certain points on the surface of
catalysts (p.187) at which the chemical reactions are
initiated or take place. Not all the surface is catalytically
active, and some catalysts are specially treated prior to
use to increase the number of active centres.

homogeneous catalysis a catalyzed reaction in which
the catalyst (p.187) and reactants are in the same
phase, e.g. the inversion of glucose in aqueous solution
catalyzed by hydrogen ions.

heterogeneous catalysis a catalyzed reaction in which
the catalyst (p.187) and reactants form different
phases, e.g. many industrial processes involve passing
gases or vapours over solid metal catalysts. The
catalytic oxidation of ammonia to produce nitrogen
monoxide for the manufacture of nitric acid involves the
use of a solid platinum-rhodium gauze as a
heterogeneous catalyst.

$$CH_3CH_2OH \xrightarrow[\text{400°C}]{\text{Ni catalyst}} CH_3CHO + H_2$$

**heterogeneous catalytic
dehydrogenation** of ethanol
when passed over heated
nickel

substrate (n) refers to a chemical upon which a reaction
is carried out, e.g. a compound which is adsorbed
(p.178) on the active centres ($\uparrow$) of a catalyst (p.187)
before reacting further; a chemical compound in which
one functional group (p.36) is substituted for another.

acid/base catalysis homogeneous catalysis ($\uparrow$), in
solution, in which the catalyst is either protons (giving
general acid catalysis), or hydroxyl ions (giving general
base catalysis), e.g. the hydrolysis of esters may be
catalyzed by either acid or base conditions.

electrochemical cell a cell in which a chemical reaction takes place and some of the energy is released in the form of electrical energy. The cell consists of two different metallic conductors in contact with an electrolytic conductor. When the two metallic conductors are connected, an electric current flows arising from a potential difference between them.

galvanic cell same as **electrochemical cell** (↑).

voltaic cell an electrochemical cell (↑) usually restricted to an irreversible cell (↓).

electrode potential at the surface of separation between a metallic conductor and an electrolytic conductor there exists a potential difference, called the electrode potential. The potential difference between the electrodes of an electrochemical cell (↑) is the *algebraic* difference of these electrode potentials, e.g. $-0.76V - (+0.32V) = -1.08V$.

Helmholtz double layer a metal in contact with a solution containing ions of the metal may exhibit two tendencies; (1) metal atoms may become ions, leaving the metal negatively charged; (2) metal ions may deposit metal atoms on the electrode, leaving the metal positively charged as electrons are withdrawn to convert ions to atoms. In either case a double layer of charge is formed at the electrode surface.

electrolytic solution pressure the tendency of a metal to lose ions to an electrolyte surrounding the metal. If this is greater than the deposition pressure (↓), the electrode potential (↑) is negative.

deposition pressure the tendency of metal ions from an electrolyte to deposit on a metal. If this is greater than the electrolytic solution pressure, the electrode potential (↑) is positive.

electromotive force (e.m.f.) the potential of the electrical energy from an electrochemical cell; it is measured in volts. It is the electrical force which drives current round an electrical circuit. Electromotive force is measured as the potential difference between the electrodes of a cell on open circuit, i.e. when no current flows. If current flows from the cell, the potential difference between the electrodes is less than the e.m.f.

reversible cell an electrochemical cell (↑) in which the chemical process producing the e.m.f. (↑) is a reversible reaction (p.185). At equilibrium, an external potential difference balances exactly the e.m.f. of the cell. A slight increase or decrease of this potential

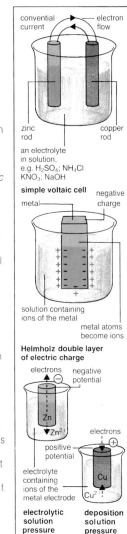

conventional current — electron flow

zinc rod — copper rod

an electrolyte in solution, e.g. H_2SO_4; NH_4Cl KNO_3; NaOH

simple voltaic cell

metal — negative charge

solution containing ions of the metal

metal atoms become ions

Helmholz double layer of electric charge

electrons — negative potential

Zn

Zn^{2+}

electrons

positive potential

electrolyte containing ions of the metal electrode

Cu

Cu^{2+}

electrolytic solution pressure	deposition solution pressure

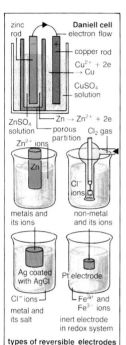

types of reversible electrodes

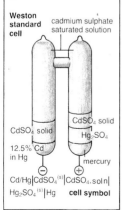

Cd/Hg|CdSO$_4$$^{(s)}$|CdSO$_4$, soln|
Hg$_2$SO$_4$$^{(s)}$|Hg **cell symbol**

difference can cause the chemical process to proceed in either direction, e.g. in a Daniell cell (↓) the reversible reaction is: $Zn + Cu^{2+} \rightleftharpoons Zn^{2+} + Cu$ (e.m.f. = 1.10 V). If the external potential difference is less than 1.10 V, the reaction goes to the right, and vice versa.

Daniell cell a reversible cell (↑) with a zinc rod in contact with a solution of zinc ions, and a copper rod in contact with a solution of copper ions. The two solutions are separated by a porous partition.

reversible electrodes the electrodes in a reversible cell (↑). There are three types; (1) a metal, or a non-metal, in contact with a solution of its own ions. (2) a metal and a sparingly soluble salt of the metal in contact with a solution of the anion of the salt, e.g. Ag; AgCl$_{(s)}$; HCl; (3) an inert electrode, e.g. Pt or Au, in contact with a solution containing both oxidized and reduced states of a redox system, e.g. Fe^{2+}/Fe^{3+}; Sn^{2+}/Sn^{4+}.

irreversible cell an electrochemical cell (↑) in which an irreversible chemical reaction takes place, e.g. a zinc rod and a copper rod in a solution of sulphuric acid. The zinc metal is changed to zinc ions, but the reverse reaction does not take place.

cell convention the electrodes and electrolytes are separated by vertical lines. Oxidation takes place at the left-hand electrode, which is also the negative pole of the cell. If two separate electrolytes are used, and they are in direct contact, a single line is used; if they are connected through a porous membrane, a dotted line is used; if they are connected by a salt bridge, a double line, ‖, is used between the symbols for the solution.

standard electromotive force the standard e.m.f. of a cell is measured at 25°C using a potentiometer, *see diagram*. An accumulator, C, forms a potential gradient along the wire AB. A point of balance, D, is found using the cell. At this point no current passes, as shown by the galvanometer. Hence the e.m.f. of the cell is equal to the potential drop, AD, along the wire. The point of balance, E, is then found using a standard cell (↓). The relationship is:

$$\frac{\text{e.m.f. of cell}}{\text{e.m.f. of standard cell}} = \frac{AD}{AE}$$

standard cell a cell whose e.m.f. (↑) and change of e.m.f. with temperature is known accurately.

Weston cell the structure of this standard cell is shown in the diagram. Its e.m.f. is 1.018636 volts at 20°C; this decreases by 4×10^{-5} V per degree rise of temperature.

half cell an electrochemical cell (p.190) can be regarded as two half cells connected by a salt bridge (p.196), a porous membrane, or by direct contact. The e.m f. of the cell is then equal to the algebraic difference of the electrode potentials (p.190) of the two electrodes, each being a reversible electrode (p.191).

hydrogen electrode a single reversible electrode consisting of a platinum foil electrode, coated with platinum black (finely divided platinum), half immersed in an acid solution, and half in pure hydrogen gas at one atmosphere pressure. The hydrogen ions in the acid solutions have unit activity. By convention, the electrode potential of the hydrogen electrode is zero, and it is known as the standard hydrogen electrode.

reference electrode the hydrogen electrode (↑) is not convenient for practical work, although it is the primary reference electrode. Secondary reference electrodes are used; these include the calomel electrode (p.133). A cell uses a reference electrode and an unknown electrode whose electrode potential is to be determined. *See also* **glass electrode** (p.133).

hydrogen scale by assigning a value of zero to a standard hydrogen electrode, all other reversible electrodes can be given a standard value. Electrode potentials based on this zero refer to the hydrogen scale.

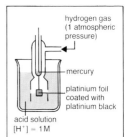

hydrogen gas
(1 atmospheric
pressure)

mercury

platinum foil
coated with
platinum black

acid solution
$[H^+] = 1M$

**simple standard
hydrogen electrode**

hydrogen scale

electrode	reaction	$E°_{(v)}$
K/K^+	$K^+ + e \rightarrow K$	-2.925
Na/Na^+	$Na^+ + e \rightarrow Na$	-2.714
Zn/Zn^{2+}	$Zn^{2+} + 2e \rightarrow Zn$	-0.763
Fe/Fe^{2+}	$Fe^{2+} + 2e \rightarrow Fe$	-0.440
Cd/Cd^{2+}	$Cd^{2+} + 2e \rightarrow Cd$	-0.408
Sn/Sn^{2+}	$Sn^{2+} + 2e \rightarrow Sn$	-0.140
Pb/Pb^{2+}	$Pb^{2+} + 2e \rightarrow Pb$	-0.126

$Pt(H_2)/H^+$	$H^+ + e \rightarrow \frac{1}{2}H_2$	0.000
Cu/Cu^{2+}	$Cu^{2+} + 2e \rightarrow Cu$	$+0.337$
$Pt(O_2)/OH^-$	$\frac{1}{2}O_2 + H_2O + 2e \rightarrow 2OH^-$	$+0.401$
$Pt (I_2)/I^-$	$\frac{1}{2}I_2 + e \rightarrow I^-$	$+0.536$
Hg/Hg_2^{2+}	$\frac{1}{2}Hg_2^{2+} + e \rightarrow Hg$	$+0.789$
Ag/Ag^+	$Ag^+ + e \rightarrow Ag$	$+0.799$
$Pt(Br_2)/Br^-$	$\frac{1}{2}Br^2 + e \rightarrow Br^-$	$+1.065$
$Pt(Cl_2)/Cl^-$	$\frac{1}{2}Cl_2 + e \rightarrow Cl^-$	$+1.359$

**standard electrode
potentials**

standard electrode potential the electrode potential depends on; (1) the concentration of ions in the electrolyte surrounding the electrode; (2) the temperature; and (3) the pressure, if the electrode is a gas, e.g. hydrogen, chlorine. The standard electrode potential is measured at a concentration of $1 \, mol \, dm^{-3}$, or for very accurate work at unit activity (p.181) for the ionic species associated with the electrode. The temperature is standardized at 25°C and the pressure at one atmosphere. The value of the potential under IUPAC rules is negative if the electrode forms the negative pole of a cell with a hydrogen electrode, e.g. the standard electrode potential of zinc is $-0.76\,V$ and of copper is $+0.34\,V$. In America, the oxidation potential is used; this reverses the sign of the standard electrode potential. The symbol for a standard electrode potential is E°. The effect of concentration (or activity) and temperature on a standard electrode potential is determined by the Nernst equation ($\downarrow$).

Nernst equation the electrode potential, E, of an element in contact with its ions in solution is related to the standard electrode potential, E°, by $E = E^\circ + RT\ln c/zF$, where R is the gas constant, T the absolute temperature, F the faraday, c the ionic concentration, and z the electrovalency of the ion. If T is $298\,K$ ($25°C$) then $E = E^\circ + 0.059 \log c/z$. For accurate calculations, concentration is replaced by the activity of the ion. The value of z is positive for cations and negative for anions, e.g. $z = +2$ for Cu^{2+} and -1 for Cl^-. If $c = 1$, i.e. a standard condition, then $\log c = 0$ and $E = E^\circ$.

concentration effect for a monoelectrovalent cation, $z = 1$ in the Nernst equation, hence dilution to $1/10$ of the concentration gives a concentration effect of -0.059 volts. E° for zinc = $-0.763\,V$ and $z = +2$; if $[Zn^{2+}] = 0.01\,M$ then
$E = -0.763 + (0.059/+2) \times (-2) = -0.822\,V$.
E° for chlorine = $+1.359\,V$ and $z = -1$; if $[Cl^-] = 0.01\,M$ then $E = 1.359 + (0.059/-1) \times (-2) = 1.477\,V$. For accurate measurements, activities of ions are used in place of concentrations. All calculations assume $T = 298\,K$.

standard cell potential this is calculated for $T = 298\,K$ from the Nernst equation ($\uparrow$). A cell is written down with the negative pole on the left-hand side. The two half cells have standard electrode potentials of E°_R on the right-hand side and E°_L on the left. The standard cell potential is: $E^\circ = E^\circ_R - Eu^\circ_L$ giving the sign of the e.m.f. as the polarity of the right-hand electrode. A positive value of E° means that the chemical reaction in the cell proceeds spontaneously from left to right. For ionic activities other than unit activity, the e.m.f. is:

$$E = E^\circ - \frac{RT}{3F}\ln\frac{a_L}{a_R}$$

where a_L is the ionic activity in the left half cell and a_R in the right half cell. If two electrodes share the same electrolyte then $a_L = a_R$, and the cell potential is independent of the concentration of the electrolyte.

equilibrium constant the equilibrium constant, K_c, is related to the standard Gibbs free energy change, ΔG°, of a reversible chemical system by: $\Delta G^\circ = -RT\ln K_c$ (*see* p.112). ΔG° is related to the standard cell potential (p.193) by $\Delta G^\circ = -nFE^\circ$. Hence:

$$E^\circ = \frac{RT}{nF}\ln K_c = \frac{0.059}{n}\log K$$

In the Daniell cell (p.191), $E^\circ = 1.100\,V$ and $K_c = 1.7 \times 10^{37}$. If zinc is placed in a solution of Cu^{2+} ions, the ratio of concentrations at equilibrium will be $[Zn^{2+}]/[Cu^{2+}] = 1.7 \times 10^{37}$, i.e. hardly any copper ions remain.

$$E = E_{Cu}^\circ - E_{Zn}^\circ$$
$$- \frac{0.059}{2}\log\frac{[Zn^{2+}]}{[Cu^{2+}]}$$
$$= 0.337 + 0.763$$
$$- \frac{0.059}{2} \times 1.7 \times 10^{37}$$
$$= 1.100 - 1.100$$
$$= 0\,v$$
at equilibrium for the reaction

oxidation-reduction electrode an inert electrode, usually platinum or gold, in contact with a solution containing both oxidized and reduced states of an ion, e.g. Fe^{2+}/Fe^{3+}; Sn^{2+}/Sn^{4+}. The electrode reaction for Fe^{2+}/Fe^{3+} is: $Fe^{3+} + e \rightarrow Fe^{2+}$, with a redox potential ($\downarrow$) of $+0.771\,V$.

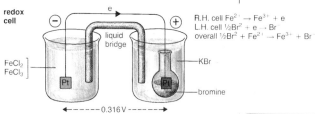

R.H. cell $Fe^{2+} \rightarrow Fe^{3+} + e$
L.H. cell $\frac{1}{2}Br_2 + e \rightarrow Br^-$
overall $\frac{1}{2}Br_2 + Fe^{2+} \rightarrow Fe^{3+} + Br^-$

redox cell a redox cell is shown in the diagram, together with the overall reaction, the e.m.f. of the cell, and the calculations of the equilibrium constant, K. Bromine oxidizes iron (II) salts to iron (III) salts and the reaction goes practically to completion. The ions, K^+ and Cl^-, play no significant part in the reaction. The liquid bridge is a salt bridge (p.196) connecting the two solutions. A cell using two oxidation-reduction electrodes is: $Pt \mid Sn^{2+}, Sn^{4+} \parallel Fe^{2+}, Fe^{3+} \mid Pt$. The overall reaction is: $Sn^{2+} + 2Fe^{3+} \rightarrow Sn^{4+} + 2Fe^{2+}$, i.e. tin (II) ions reduce iron (III) ions. The standard e.m.f. of the cell is $+0.62\,V$, so $\log K_c = (0.62 \times 2)/0.059$ and $K_c = 10^{21}$, so reduction goes to completion.

$$K = \frac{[Fe^{3+}][Br^-]}{[Fe^{2+}]}$$

$$E^\circ = \frac{0.059}{n}\log K \; (n = 1)$$

$$\log K = \frac{0.316 \times 1}{0.059} = 5.3559$$

$$K = 2.27 \times 10^5$$

redox potential the potential of a half cell (p.192) with an oxidation-reduction electrode ($\uparrow$). A standard redox potential, E°, has equal concentrations of the oxidized

$$E = \frac{0 \cdot 059}{n} \log K \; (n = 5)$$

$$\log K = \frac{0 \cdot 22 \times 5}{0 \cdot 059} = 18 \cdot 64$$

$$K = 4 \times 10^{18}$$
oxidation is complete

and reduced states of the ion. The actual potential, E, is related to the standard redox potential by:

$$E = E^{\circ} + \frac{RT}{nF} \; \ln \frac{[Oxid]}{[Red]}$$

For standard conditions (*see* **Nernst equation** p.193) this becomes:

$$E = E^{\circ} + \frac{0.059}{n} \log \frac{[Oxid]}{[Red]}$$

where [Oxid] is the concentration of the oxidized species and [Red] the concentration of the reduced species. For the Fe^{2+}/Fe^{3+} electrode, $E^{\circ} = 0.771\,V$.

If $[Fe^{3+}] = [Fe^{2+}]$, then $\log \frac{[Fe^{3+}]}{[Fe^{2+}]} = 0$ and $E = E^{\circ}$.

If $[Fe^{3+}]/[Fe^{2+}] = 10$ then $E = +0.771 + \frac{0.059 \times 1}{1}$

($n = 1$, $\log 10 = 1$). So $E = 0.830\,V$.

redox pH effect oxidizing agents that are used in the presence of acids, or alkalis, have redox potentials (↑) that are dependent on the pH (p.132) of the solution, as well as the concentration of the oxidized and reduced states of the agent. e.g. the manganate (VII) ion reacts as: $MnO_4^- + 8H^+ + 5e \rightarrow Mn^{2+} + 4H_2O$.
The redox potential is:

$$E = E^{\circ} + \frac{RT}{5F} \; \ln \frac{[MnO_4^-][H^+]^8}{[Mn^{2+}]}$$

under standard conditions [Oxid] = [Red] and $[H^+] = 1$, i.e. pH = 0.

$$E = E^{\circ} + \frac{0.059}{5} \log \frac{[MnO_4^-]}{[Mn^{2+}]} + \frac{0.059 \times 8}{5} \; \log [H^+]$$

but $\log [H^+] = -pH$, so the last term becomes $(-0.0944 \times pH)$. $E^{\circ} = +1.52\,V$ for this electrode. At pH = 5 and $[MnO_4^-] = [Mn^{2+}]$, $E = +1.52 - 0.47 = +1.05\,V$.

redox pH effect

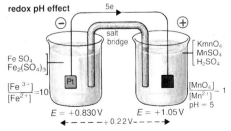

standard redox potential see **redox potential** (p.194) and **redox pH effect** (p.195).

liquid junction potential if two solutions are in contact, a liquid junction potential exists between them. This potential can be calculated, but it is preferable to eliminate it.

salt bridge a tube containing a saturated solution of potassium chloride solution connecting two solutions in an electrochemical cell (p.190). Potassium ions and chloride ions have almost identical ionic mobilities so positive and negative charge is transferred equally between the two solutions; this minimizes the liquid junction potential.

concentration cell a cell composed of two identical electrodes each in a solution of different concentrations. The concentration effect (p.193) gives rise to an e.m.f.; the value of the e.m.f. can be calculated by applying the Nernst equation (p.193). A concentration cell with two copper electrodes is shown in the diagram. If the electrodes are connected the reaction proceeds until the concentrations are equal in both half cells.

electrochemical series standard electrode potentials and redox potentials can be put in a list, the electrochemical series. An oxidizing agent will oxidize any element or reducing agent higher on the list and a reducing agent will reduce any element or oxidizing agent lower on the list. Some typical potentials are shown in the list; the values use the hydrogen scale (p.192). A difference of 0.4 V between standard electrode potentials, where 1 faraday is transferred in a reaction, leads to an equilibrium constant 6×10^6, i.e. the reaction goes to completion. The electrochemical series is frequently restricted to a list of elements in the order of their standard electrode potentials.

standard oxidation potential the standard electrode potential (p.192) under IUPAC rules is a reduction potential. In the electrochemical series (↑) under IUPAC rules the electrode potentials are in descending order of strength of reducing agents. The standard oxidation potential of any electrode is equal to the standard electrode potential with the sign reversed, e.g. the standard oxidation potential of zinc is $+0.771$ V.

free energy diagram the Gibbs free energy (p.112) for an element is zero; the change in free energy, ΔG°, can be calculated for an ion from the standard electrode potential, using the relation $\Delta G = -nFE^0$.

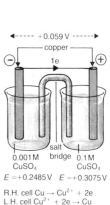

$E = +0.2485\,\text{V}$ $E = +0.3075\,\text{V}$

R.H. cell $Cu \rightarrow Cu^{2+} + 2e$
L.H. cell $Cu^{2+} + 2e \rightarrow Cu$

concentration cell

reaction	E° (v)
$Li^+ + e \rightarrow Li$	-3.03
$Zn^{2+} + 2e \rightarrow Zn$	-0.763
$Fe^{2+} + 2e \rightarrow Fe$	-0.44
$Cu^{2+} + 2e \rightarrow Cu$	$+0.337$

← increasing strength reducing agents

← increasing strength oxidizing agents

electrochemical series

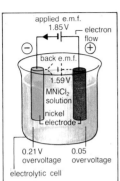

applied e.m.f.
1.85 V
electron flow
back e.m.f.
1.59 V
$MNiCl_2$ solution
nickel electrode
0.21 V overvoltage
0.05 overvoltage

electrolytic cell

electrolytic process

electrode	over-voltage
Pt black	0.01V
Pt smooth	0.26V
Fe	0.40V
Ag	0.46V
Cu	0.54V
Sn	0.85V
Pb	0.88V
Hg	1.04V

hydrogen overvoltages in M hydrochloric acid and 1 mA cm^{-2}

electrode	over-voltage

oxygen overvoltages M KOH solution and 0.1 mA cm^{-2}

entropy measurement the Gibbs-Helmholtz equation (p.114) shows that: $\Delta S = -(\delta G/\delta T)$. $\Delta G = -nFE^{\circ}$, *see* **free energy diagram** (↑), so $\Delta S = nF(\delta E/\delta T)$ allowing ΔS to be calculated. In the following cell:

$$Ag \mid AgCl_{(s)} \mid KCl_{(aq)} \mid Hg_2Cl_{2(s)} \mid Hg$$

the reaction is $Ag + \frac{1}{2}Hg_2Cl_2 \rightarrow AgCl + Hg$. E° is +0.0455 V, and the temperature coefficient of the cell is $+3.38 \times 10^{-4}\,V\,K^{-1}$. One electron is transferred in the reaction so $n = 1$, $F = 96\,500$ coulomb,
$$dE/dT = +3.38 \times 10^{-4}.$$
$$\Delta S = 1 \times 96\,500 \times 3.38 \times 10^{-4} = 32.6\,JK^{-1}$$
which is the standard entropy change for the reaction.

electrolytic process the passage of a current through an electrolyte to effect its decomposition. The applied voltage necessary to cause electrolysis depends on the reversible electrode potentials of the electrodes and on overvoltage (↓), e.g. in the electrolysis of M nickel chloride solution, using nickel electrodes, the decomposition voltage (p.126) is 1.85 V; the reversible electrode potentials of nickel and chlorine, in a molar solution, are −0.23 V and +1.36 V respectively, producing a cell e.m.f. of 1.59 V; the difference, 0.26 V, is the overvoltage of the electrodes. For an electrolytic process, the full equation is: (applied voltage) = (cell voltage) + (overvoltage) + (current × resistance).

electrolytic cell a cell in which an electrolytic process (↑) takes place.

overvoltage (*n*) the excess voltage over and above the reversible electrode potential before electrolysis commences. Overvoltages occur when gases are liberated at an electrode. In the electrolysis of aqueous solutions, hydrogen and oxygen gases may be liberated. These form hydrogen and oxygen electrodes which produce an e.m.f. in opposition to the applied voltage. The nature of the metal affects the potential of such gaseous electrodes. Similarly, over-voltages may occur with other gases. The magnitude of an overvoltage: (1) decreases with a rise in temperature of the electrolyte; (2) increases with the current density.

overpotential (*n*) another name for **overvoltage** (↑).

back e.m.f. the liberation of products during electrolysis forms an electrochemical cell (p.190) or the electrodes themselves form a cell, and the cell has an e.m.f. which acts in opposition to the applied voltage. The cell produces a back e.m.f.

theoretical decomposition voltage the voltage equal to the back e.m.f. (p.197) in electrolysis. Once the applied voltage exceeds the back e.m.f., electrolysis should take place. If E_A is the electrode potential (p.190) of the anode, and E_C the electrode potential of the cathode, then: theoretical decomposition voltage = $E_A - E_C$ (algebraic difference).

experimental decomposition voltage this is measured by the circuit shown in the diagram. A graph of the results is used to determine the experimental decomposition voltage. If E_T is the theoretical decomposition voltage, ω_A is the overvoltage (p.197) at the anode and ω_C is the overvoltage at the cathode, then: experimental decomposition voltage = $E_T + \omega_A + \omega_C$.

discharge potential the voltage required to produce continuous discharge of ions at an electrode. If E is the electrode potential (p.190) for an ionic species, ω the overvoltage at the electrode, then: discharge potential = $E + \omega$, e.g. in the electrolysis of M NaCl solution with a steel cathode, the electrode potential of hydrogen is $E_H = 0 + 0.059 \log 10^{-7}$ (pH = 7 for the solution). The overvoltage of hydrogen on the cathode is $-0.2\,V$. So the discharge potential of hydrogen is $0 - 0.41 - 0.2 = -0.61\,V$.

selective discharge if two ionic species of the same electrical sign can be discharged at an electrode, then the ion with the lower discharge potential (↑) will be liberated, e.g. in the electrolysis of M NaCl solution with a graphite anode, the electrode potential of oxygen is: $E_O = +0.40 - 0.059 \log 10^{-7} = +0.81\,V$ (pH = 7 for the solution). The electrode potential of chlorine = $+1.36\,V$ (standard conditions, no overvoltage). The overvoltage for oxygen on graphite is $+0.6\,V$, so the discharge potential for hydroxyl ions is $+1.41\,V$. As $1.36\,V$ is lower than $1.41\,V$, chlorine is discharged preferentially to oxygen.

decomposition process for solutions of approximately equal concentrations of metallic ions, the metal with the lowest electrode potential will be deposited in electrolysis, e.g. copper is deposited before zinc. The decomposition potential is the discharge potential (↑) of the metal, which is usually equal to its electrode potential. For two metals with deposition potentials of approximately the same magnitude, both metals are deposited as an alloy. With a mercury cathode, a discharge metal usually forms an amalgam.

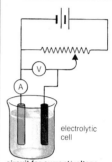

circuit for current/voltage relation in an electrolytic process

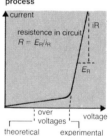

decomposition voltage

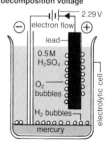

H_2 overvoltage $0.78\,V$
O_2 overvoltage $0.31\,V$

cathode potential = $0 - 0.78 = -0.78\,V$
anode potential = $+1.2 + 0.31 = +1.51\,V$
decomposition voltage = $1.51 - (-0.78) = 2.29\,V$

The deposition potential of a metal increases as concentration of its ions falls.

polarization (n) (1) during the electrolysis of an electrolyte, metals may be deposited or gases liberated. The electrodes acquire a potential from such products and the reversible electrode potential is changed, the electrode becomes an irreversible electrode. The process is called polarization and the electrodes are polarized; (2) in a voltaic cell (p.190), the electrodes may electrolyze the electrolyte and produce polarization, making the cell irreversible. In aqueous solutions, polarization is caused by hydrogen bubbles on the cathode and oxygen bubbles on the anode.

polarization voltage the polarization voltage is the difference between the experimental decomposition voltage and the e.m.f. of a reversible cell for the same chemical process.

concentration polarization a decrease in the concentration of ions in the solution immediately surrounding an electrode raises the discharge potential (↑). This is concentration polarization and it is overcome by vigorous stirring of the electrolyte.

depolarizer (n) when hydrogen is liberated, polarization (↑) can be prevented by adding an oxidizing agent to the electrolyte. When oxygen is liberated, polarization can be prevented by adding a reducing agent. Depolarizers are used in voltaic cells (p.190).

redox reaction a reaction involving oxidation and reduction, as both processes occur together. A redox reaction with ionic species involves the transfer of electrons, e.g. Fe^{3+} ions are reduced by I^- ions to Fe^{2+} ions, and I^- ions are oxidized by Fe^{3+} ions to I_2 atoms and molecules. The reaction is shown by a redox cell. Some ionic redox reactions need the presence of hydrogen or hydroxyl ions.

oxidation potential the potential of a reversible electrode involving an oxidation process, e.g. $Na \rightarrow Na^+ + e$; $Cl^- \rightarrow \frac{1}{2}Cl_2 + e$. The standard oxidation potential of Na/Na^+ is 2.71 V and of $Pt \,|\, Cl_{2_{(g)}} \,|\, Cl^-$ is -1.36 V. The magnitudes are those of the standard electrode potentials with the sign reversed.

reduction potential the potential of a reversible electrode involving a reduction process, e.g. $Na^+ + e \rightarrow Na$; $\frac{1}{2}Cl_2 + e \rightarrow Cl^-$. These potentials are the same as standard electrode potentials (p.192).

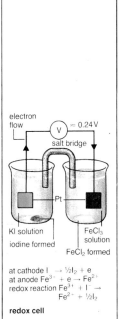

electron flow
$\approx 0.24\,V$
salt bridge
Pt
KI solution
iodine formed
$FeCl_3$ solution
$FeCl_2$ formed

at cathode $I \rightarrow \frac{1}{2}I_2 + e$
at anode $Fe^{3+} + e \rightarrow Fe^{2+}$
redox reaction $Fe^{3+} + I^- \rightarrow Fe^{2+} + \frac{1}{2}I_2$

redox cell

oxidation state the oxidation state of elements in ionic
compounds is equal to the charge on the ions, e.g. in
Na^+Cl^- the oxidation state of sodium is $+1$ and of
chlorine is -1. The oxidation state of free or uncombined
elements is zero. Oxidation is an increase in oxidation
state, e.g Fe^{2+} to Fe^{3+} an increase from $+2$ to $+3$; Cl^-
to $\frac{1}{2}Cl_2$ an increase from -1 to 0. Oxidation states of
metals are indicated by oxidation numbers ($\downarrow$).

oxidation state and number

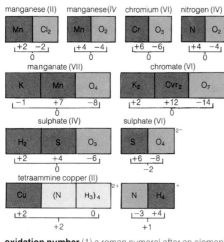

oxidation number (1) a roman numeral after an element
showing its oxidation state, e.g. tin (II) chloride and
tin (IV) chloride showing oxidation states of $+2$ and $+4$
respectively. (2) a number allotted to an element in a
covalent compound, an anion, or a complex ion, by
considering all bonds to be ionic instead of covalent.
The algebraic sum of all oxidation numbers in an
uncharged compound is zero; in an ion it is equal to the
overall charge on the ion. *See also* **coordination
number** (p.169).

disproportionation (*n*) if an element can exist in more
than one oxidation state ($\uparrow$), and its ion undergoes both
oxidation and reduction at the same time, the process is
called disproportionation, e.g. a copper (I) ion in
aqueous solution undergoes disproportionation thus:
$2Cu^+_{(aq)} \rightarrow Cu + Cu^{2+}_{(aq)}$. The standard electrode
potentials, *see diagram*, show that Cu/Cu^{2+} is a better
reducing agent than Cu/Cu^+, and that Cu^{2+}/Cu is a

disproportionation of Cu^+

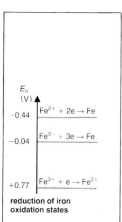

E_0
(V)

-0.44 $Fe^{2+} + 2e \rightarrow Fe$

-0.04 $Fe^{3+} + 3e \rightarrow Fe$

$+0.77$ $Fe^{3+} + e \rightarrow Fe^{2+}$

**reduction of iron
oxidation states**

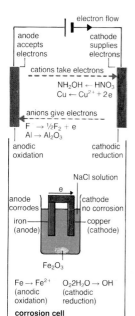

electron flow

anode
accepts
electrons

cathode
supplies
electrons

cations take electrons

$NH_2OH \leftarrow HNO_3$
$Cu \leftarrow Cu^{2+} + 2e$

anions give electrons

$F^- \rightarrow \frac{1}{2}F_2 + e$
$Al \rightarrow Al_2O_3$

anodic
oxidation

cathodic
reduction

NaCl solution

e

anode
corrodes

cathode
no corrosion

iron
(anode)

copper
(cathode)

Fe_2O_3

$Fe \rightarrow Fe^{2+}$
(anodic
oxidation)

$O_2 2H_2O \rightarrow OH$
(cathodic
reduction)

corrosion cell

better oxidizing agent than Cu^+/Cu. The reactions are:

oxidizing action $Cu^+ + e \rightarrow Cu$
(Cu$^+$ accepts electrons)

reducing action $Cu^+ \rightarrow Cu^{2+} + e$
(Cu$^+$ donates electrons)

Copper (I) compounds are stable only if they are insoluble in water, or if they form complex compounds. e.g. $[Cu(CN)_4]^{3-}$ tetracyanocuprate (I). The standard electrode potentials for the oxidation states of iron are also given for comparison.

anodic oxidation during electrolysis, an anode attracts electrons, and behaves as an oxidizing agent, e.g. $F^- \rightarrow \frac{1}{2}F_2 + e$, a reaction that cannot be carried out any other way. Some metal anodes are oxidized by electrolysis, e.g. an aluminium anode has a thin film of oxide thickened by electrolysis; the process is called anodising (↓). Anodic oxidation of sulphuric acid yields perdisulphuric acid $H_2S_2O_8$.

cathodic reduction during electrolysis, a cathode supplies electrons and behaves as a reducing agent. A lead cathode has a high overvoltage (p.197) for hydrogen, and this can reduce molecules not necessarily involved in the electrolytic process, e.g. nitric acid is reduced to hydroxylamine NH_2OH, nitrobenzene is reduced to phenylamine (aniline).

anodising (*n*) the anodised film on aluminium can absorb dyes during the electrolysis. Anodised aluminium is resistant to corrosion (↓).

corrosion (*n*) the action of atmospheric oxygen, also other gases, on metals to form a layer of oxide, or other compounds. **Dry corrosion** involves atmospheric gases only. **Wet corrosion** involves water in addition to atmospheric gases. Any two different metals in an aqueous solution exhibit corrosion. A corrosion cell has iron forming iron (II) ions, an anodic oxidation process, and oxygen reduced to hydroxyl ions, a cathodic reduction. In the corrosion cell, iron is considered to be the anode and copper the cathode. Any two metals can similarly form a corrosion cell with the nobler metal (the one with the more positive electrode potential) forming the cathode, and the less noble metal forming the anode. The anode corrodes, and the cathode reduces oxygen thus: $O_2 + 2H_2O + 4e \rightarrow 4OH^-$. The cathodic reduction can be demonstrated by **differential aeration** *see diagram*. Whichever iron electrode is aerated becomes the cathode. **corrode** (*v*).

rusting (*n*) the corrosion (p.201) of iron. Iron as a single
metal corrodes. Atmospheric oxygen forms an oxide
film of varying thickness. Where the film is thin, or
cracked, iron (II) ions enter any aqueous solution
covering the iron, and this area acts as an anode. The
oxide layer acts as a cathode *see diagram*, and further
rust is deposited; the reaction is:

$$Fe^{2+} + 2OH^- \rightarrow Fe(OH)_2 \xrightarrow{[O_2]} Fe_2O_3$$

Iron can be protected by a thin layer of another metal,
e.g. zinc or tin. If this thin layer is scratched, or cracked,
a corrosion cell is formed. Zinc ions enter any aqueous
solution, forming zinc hydroxide which seals the crack.
Tin acts as a cathode, making iron the anode, the iron
rusts, and expands the crack. **rust** (*n*), **rust** (*v*).

bimetallic couple any pair of different metals which,
when immersed in a solution of an electrolyte, form an
electrochemical cell, and produce galvanic corrosion.

cathodic protection a zinc coating on iron provides
cathodic protection, as the iron cathode does not rust;
zinc is a **sacrificial coating**. Cathodic protection is
given to iron structures in sea water by connecting them
to a magnesium rod. The magnesium gradually dissolves,
forming magnesium ions, as it acts as the anode; this
action protects the cathode which does not rust.

tarnish (*n*) the corrosion (p.201) on a lustrous surface,
e.g. silver and copper form tarnish, but gold and
platinum do not. The tarnish is a thin film of oxide or
sulphide. **tarnish** (*v*).

passivity (*n*) the action of concentrated nitric acid on
some metals, e.g. iron, cobalt, nickel, chromium, forms
a thin layer of oxide, giving the metal passivity, i.e. it no
longer corrodes, nor is it attacked by acids and other
corrosive substances. Passive iron has a high positive
electrode potential. **passive** (*adj*).

galvanic corrosion corrosion taking place when two
different metals are in contact with each other and an
aqueous solution. *See* **bimetallic couple** (↑).

differential aeration on a metal surface in contact with
an aqueous solution of oxygen, the areas where the
oxygen concentration is low will corrode. This is the
differential aeration principle. Electric currents flow
between the well-aerated and the poorly-aerated
surfaces. Differential aeration explains crevice
corrosion. *See diagram*.

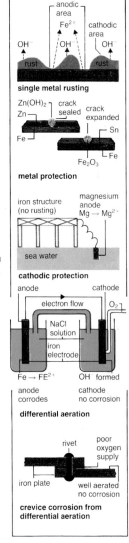

single metal rusting

metal protection

cathodic protection

differential aeration

**crevice corrosion from
differential aeration**

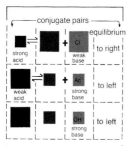

Brønsted-Lowry theory

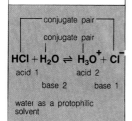

water as a protophilic solvent

protophilic solvent

Brønsted-Lowry theory an acid is a substance that tends to give protons to a base, and a base is a substance that tends to accept protons from an acid. An acid is a proton-donor ($\downarrow$) and a base is a proton-acceptor ($\downarrow$) summarized as:

$$A \rightleftharpoons H^+ + B$$
acid   proton   base

proton-donor an acid in the Brønsted-Lowry theory ($\uparrow$). The strength of an acid is measured by its ability to give protons to a base.

proton-acceptor a base in the Brønsted-Lowry theory ($\uparrow$). The strength of a base is measured by its ability to accept protons.

conjugate pair in the relation: $A \rightleftharpoons H^+ + B$, the acid, A, and base, B, which differ by a proton, are a conjugate pair. Every acid has its conjugate base, and every base its conjugate acid by the Brønsted-Lowry theory ($\uparrow$).

conjugate acid if a base is strong, it accepts protons to move the equilibrium to the left in the relation of a conjugate pair, so the conjugate acid of a strong base is a weak acid, and vice versa. See **conjugate pair** ($\uparrow$).

conjugate base if an acid is strong, it donates protons to move the equilibrium to the right in the relation of a conjugate pair, so the conjugate base of a strong acid is a weak base, and vice versa. See **conjugate pair** ($\uparrow$).

protophilic (adj) describes a substance which tends to accept protons.

protophilic solvent a solvent which is a proton acceptor, i.e. a basic substance. Water is the commonest protophilic solvent. Glacial ethanoic acid is less protophilic and liquid ammonia more protophilic than water as a solvent. The protophilic nature of a solvent determines the extent of ionization of an acid; the more protophilic a solvent, the greater is the extent of ionization.

aprotic solvent a solvent which does not accept protons at all, e.g. a liquid hydrocarbon such as methyl benzene. In such a solvent even a strong acid, e.g. hydrogen chloride, does not exhibit any acidic properties.

solvent solvent effect	solvent		
	protophilic	aprotic	protogenic
strength of acid	increased	nil	decreased
strength of base	decreased	nil	increased

protogenic solvent a solvent which is a proton-donor, i.e. an acidic substance. Examples are liquid hydrogen fluoride, liquid hydrogen chloride, concentrated sulphuric acid. In such solutions, an acid has no acidic properties and all but the strongest acids act as bases, e.g. nitric acid is made to accept protons in liquid hydrogen fluoride, and in concentrated sulphuric acid: $HNO_3 + 2H_2SO_4 \rightleftharpoons NO_2^+ + 2HSO_4^- + H_3O^+$. Water acts both as a protogenic and a protophilic solvent.

neutralization (*n*) the process of making a neutral solution by the reaction between one mole of hydrogen ions and one mole of hydroxyl ions, or amounts of these ions in the same proportion. The products are not necessarily a salt and water, but are always the conjugate base (p.203) and acid respectively of the reacting acid and base, as in these examples:
$HCl + NH_3 \rightleftharpoons NH_4^+ + Cl^-$
$NH_4^+ + OH^- \rightleftharpoons H_2O + NH_3$
In a neutral solution the concentration of hydrogen ions and hydroxyl ions is equal; pH = 7 for the solution.

heat of neutralization the neutralization (↑) of any strong acid by any strong base is represented by:
$H_3O^+ + OH^- \rightleftharpoons H_2O + H_2O$
The heat of neutralization of all strong acids by all strong bases is $\Delta H = -57.3\,kJ\,mol^{-1}$ at 20°C. The heat of ionization (p.125) of water is $\Delta H = +56.7\,kJ\,mol^{-1}$, a very close agreement. The heat of neutralization of weak acids and bases is generally lower than for strong acids and bases, since the heats of ionization of the reactants have to be taken into account.

reactants	acid	base	ΔH		reactants	acid	base	ΔH
HCl and NaOH	strong	strong	−57·9		HCl and NH_3	strong	weak	−53·4
HNO_3 and NaOH	strong	strong	−57·6		CH_3COOH and NH_3	weak	weak	−50·4
CH_3COOH and NaOH	weak	strong	−56·1		HCl and ½MgO	strong	solid	−73·0

incomplete neutralization the neutralization of an acid can be represented by: $HA + B \rightleftharpoons BH^+ + A^-$. These conjugate acids and bases may establish proton transfer with water:
$BH^+ + H_2O \rightleftharpoons H_3O^+ + B$
$A^- + H_2O \rightleftharpoons OH^- + HA$
In the first reaction, the free base is regenerated, and in the second, the free acid. Either reaction opposes neutralization, and the incomplete neutralization from the reaction with water is hydrolysis.

---

conjugate pair

conjugate pair

$$NH_3 + H_2O \rightleftharpoons OH^- + NH_4^+$$

base 1 · acid 2 · base 2 · acid 1

water as a protogenic solvent

protogenic solvent

heat of neutralization

heats of neutralization at 25°C

$\Delta H\,kJ\,mol^{-1}$

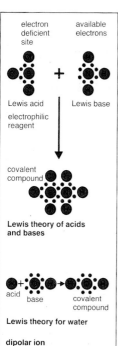

electron deficient site | available electrons

Lewis acid electrophilic reagent | Lewis base

covalent compound

Lewis theory of acids and bases

acid | base | covalent compound

Lewis theory for water

dipolar ion

+ H_2O
acid 1 | base 2

+ $^+NH_3RCOO^-$
acid 2 | base 1

dipolar ion zwitterion

+ H_2O
acid 1 | base 2

+ NH_2RCOO^-
acid 2 | base 1

Lewis theory an acid has a molecule or ion that can accept a pair of electrons to form a covalent bond. A base has a molecule or ion that can donate a pair of electrons to form a covalent bond. This is a much broader concept of acids and bases than that of the Brønsted-Lowry theory (p.203).

electron acceptor an acid under the Lewis theory (↑). A molecule or ion capable of forming a coordinate bond (p.77) with unshared electron pairs.

electron donor a base under the Lewis theory (↑). A molecule or ion which has one or more unshared electron pairs available for forming coordinate bonds.

substrate (*n*) a substance attacked by a reagent (↓).

reagent (*n*) a substance which, when added to another substance, causes a chemical reaction. There are two types of reagent: electrophilic (↓) and nucleophilic (↓).

electrophilic reagent a reagent (↑) that attacks a substance where electrons are readily available. Acids and oxidizing agents are electrophilic reagents.

nucleophilic reagent a reagent (↑) that attacks a substance which has an electron-deficient site. Bases and reducing agents are nucleophilic reagents.

electrophile (*n*) an electrophilic reagent (↑).

nucleophile (*n*) a nucleophilic reagent (↑).

monoprotic acid an acid which has one proton that can be lost from a molecule, e.g. hydrogen chloride, nitric acid, ethanoic acid.

diprotic acid an acid which has two protons that can be lost from a molecule, e.g. sulphuric acid, carbonic acid, ethanedioic acid.

aprotic acid an acid which has no proton to be lost from a molecule. Such acids are Lewis acids, as described under Lewis theory (↑), e.g. boron trichloride BCl_3 is a Lewis acid and also an aprotic acid.

amphiprotic acid an acid which has both acidic and basic characteristics, i.e. it can lose or gain protons.

dipolar ion an ion which possesses both acidic and basic groups and can be neutralized by a base or by an acid. Aminoethanoic acid, NH_2CH_2COOH, is an amino acid with a dipolar ion: $^+NH_3CH_2COO^-$. A dipolar ion has a net electric charge of zero, but its properties differ from those of uncharged, neutral molecules. In acid solution, the ion $^+NH_3CH_2COOH$ is formed; in alkaline solution, the ion $NH_2.CH_2COO^-$ is formed.

zwitterion (*n*) a dipolar ion (↑).

isoelectric point an amphiprotic acid, or electrolyte, produces dipolar ions (p.205), which can be converted to positive or negative ions. The ionic component can thus range from all positive ions at high hydrogen ion concentrations to all negative ions at low hydrogen ion concentrations. There is an intermediate condition when both ions are present in equal amounts; this is the isoelectric point.

isoelectric solution a solution of an amphiprotic acid (p.205) at its isoelectric point (↑).

strong acid the strength of an acid depends on its ability to lose a proton, and on the nature of the solvent. The strength of an acid is generally described with water as the solvent. A strong acid is a strong electrolyte (p.124) in water, and ionization is nearly complete. *See* **protophilic** and **protogenic solvents** (p.204).

strong base the strength of a base depends on its ability to gain a proton, and this also depends on the nature of the solvent. The strength of a base is generally described with water as the solvent. A strong base is a strong electrolyte (p.124).

weak acid a weak acid is a weak electrolyte (p.124).
weak base a weak base is a weak electrolyte (p.124).
inorganic reaction types reactions are homogenous (p.161) or heterogenous (p.161); they are reversible (p.185) or irreversible changes (p.109), or they can consist of consecutive reactions. A chemical change is brought about by molecular collisions, or by the inter-action of ions or ionic complexes. The reaction types are based on the mechanisms by which the reactions take place; experimental evidence does not always offer a clear explanation of a mechanism, as a reaction may appear simple but can have a complicated reaction path.

gas-phase reaction a homogenous reaction between molecules with suitable collisions providing the mechanism for molecules to react. Reactions are often found to undergo an increase in the order of reaction when the pressure is increased. An activated complex (p.184) is an intermediate compound between reactants and products in such reactions.

chain reaction[2] a gas-phase reaction (↑). *See* **chain reaction[1]** (p.187).

acid-base reaction a homogenous reaction in an aqueous solution between ions; the reaction is fast. Brønsted-Lowry acids and bases are included in such reactions. These reactions are diffusion controlled.

acid-base reaction

acid-base reactions

$$H^+ + OH^- \rightleftharpoons H_2O$$

$$H^+ + HSO_4^- \rightleftharpoons H_2SO_4$$

$$H^+ + SO_4^{2-} \rightleftharpoons HSO_4^-$$

$$H^+ + HS^- \rightleftharpoons H_2S$$

$$H^+ + NH_3 \rightleftharpoons NH_4^+$$

$$NH_4^+ + OH^- \rightleftharpoons NH_3 + H_2O$$

$$(CH_3)_2 NH_2^+ + OH^-$$
$$\rightleftharpoons (CH_3)_2 NH + H_2O$$

—— covalent bond
– – – – coordinate bond
– – – – hydrogen bond

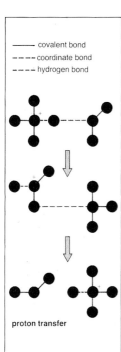

proton transfer

proton exchange a process in which a proton migrates through the hydrogen-bonded structure of water. It is not necessarily the same proton that jumps from one molecule to the next.

oxidation-reduction reaction a reaction in which one or two electrons are transferred from one ion or ionic complex to another. The transfer is from a reducing agent to an oxidizing agent. Some reactions are diffusion-controlled and thus very fast, while others form an activated complex (p.184) and are slow. Oxidation-reduction actions include one-electron, two-electron, noncomplementary and free radical reactions (↓).

one-electron reactions the oxidation-reduction reactions of the transitional elements and their complex ions. There are two possible mechanisms: outer- and inner-sphere reactions. In outer-sphere reactions, the coordination electron shells are not affected, the single electron being transferred through ligands; both reactants have stable complexes which do not undergo substitution. Inner-sphere reactions take place through the formation of an activated complex, with one ligand common to the reactants; only chromium (II) complexes definitely follow this mechanism.

two-electron reactions both metals and non-metals take part in two-electron reactions. The process is associated with a definite change in covalent coordination bonding.

homolytic reaction a reaction in which the electrons of a covalent bond, which is broken, are shared equally between the two products. Also a reaction in which species with an odd number of electrons, such as free radicals (↓) react to form a covalent bond.

free radical if a compound A:B has two groups A and B combined by a covalent bond, then the homolytic fission (p.214) of the compound by thermal or photochemical decomposition forms two free radicals, A· and B·, each retaining one electron from one bond. Free radicals are highly reactive and have a very short life, e.g. 10^{-2} seconds; they are frequently met as transient intermediates in many chemical reactions.

heterolytic reaction a reaction in which the electrons of a covalent bond, which is broken, are shared unequally between the two products. Also a reaction when two molecules, or ions, one with a lone pair of electrons, react to form a coordinate bond.

chelation (*n*) some ligands (p.78) possess more than one atom with a lone pair of electrons, e.g. bidentate, terdentate complex ions (p.144). These ligands form chelate compounds in the process of chelation. *See* **chelate ligand** (↓). **chelate** (*adj*).

complexone (*n*) a chemical compound capable of forming a complex with metal ions due to the creation of chelate ligands (↓) between itself and the ions. The best known complexone is ethylenediaminetetra-acetic acid (EDTA).

chelate ligand refers to the groups of a complexone (↑) which are joined to the metal ion in the formation of a complex compound. Ligands joined to a metal ion only at one point are monodentate ligands and those joined at two points are bidentate ligands.

EDTA *see* **complexone** (↑).

sequestering (*n*) the use of complexones (↑) to remove metal ions from solution by converting them into inactive complex compounds. The process is often used to render poisonous metals harmless and to remove metal ion impurities. **sequester** (*v*).

substitution[1] (*n*) the replacement of one ligand (p.78) by another in a complex ion. *See* **substitution**[2] (p.210).

inert complex a complex ion in which substitution of a ligand takes place slowly, i.e. over a period which can be measured in minutes.

labile complex a complex ion in which substitution of a ligand takes place very quickly, i.e. over a period measured in fractions of a second.

hydrolysis (*n*) a reaction of a complex ion in which a ligand (p.78) is replaced by a molecule of water.

acid hydrolysis hydrolysis (↑) in an acid medium; a ligand is replaced by a water molecule.

base hydrolysis hydrolysis (↑) in an alkaline medium; a ligand is replaced by a hydroxyl group.

anation (*n*) the reverse process to hydrolysis (↑); a water molecule is replaced by a ligand, e.g.
$$[Cr(H_2O)_6]^{3+} + CNS \rightarrow [Cr(H_2O)_5(CNS)]^{2+} + H_2O$$

inductomeric effect the production of a changed distribution of electrons in an organic molecule, similar to the inductive effect (p.75), which may result from the close approach of a reagent molecule, or appear in a transition stage between reactants and products. The effect can be looked upon as the polarizability of a molecule and not a permanent polarization (p.76). The electron distribution reverts to the ground state of the

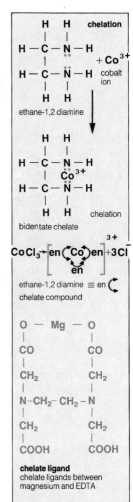

ethane-1,2 diamine

bidentate chelate

$$CoCl_3 \rightarrow [en(Co)en] + 3Cl^-$$

ethane-1,2 diamine ≡ en

chelate compound

chelate ligand
chelate ligands between magnesium and EDTA

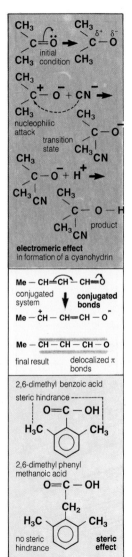

electromeric effect
in formation of a cyanohydrin

2,6-dimethyl benzoic acid
steric hindrance

2,6-dimethyl phenyl
methanoic acid

no steric hindrance / steric effect

molecule if the reagent is removed, or if a transition stage decomposes to form the original reactant. The inductomeric effect is a temporary, time-variable effect.

electromeric effect the production of a changed distribution of electrons in an organic molecule, similar to the mesomeric effect (p.77) which may result from the close approach of a reagent molecule, or appear in a transition stage between reactants and products. The effect can be looked upon as the polarizability of a molecule and not a permanent polarization (p.76). The electron distribution reverts to the ground state of the molecule if the reagent is removed, or if a transition stage decomposes to form the original reactant. The electromeric effect is a temporary, time-variable effect.

electron-withdrawing group a group of atoms, in an organic molecule, which attracts electrons from the remaining atoms in the molecule. Examples are: $-CN$; $-NO_2$; $=C=O$; $-COOR$; $-NR_3$. The group produces an inductive effect or a mesomeric effect (p.77).

electron-donating group a group of atoms, in an organic molecule, which gives electrons to the remaining atoms in the molecule. Examples are: $-CH_3$; $-OH$; $\equiv N$; CN^-. The group produces an inductive effect or a mesomeric effect (p.77).

conjugated bonds a mesomeric effect (p.77) is transmitted down a conjugative system (p.76), as shown in the diagram, of a molecule with a carbonyl group. Delocalization of the π-bonds takes place, so that electron-rich atoms result at each carbon atom in the conjugated chain. The effect does not become weaker away from the carbonyl group, or any group producing delocalization.

steric effects the size of organic groups in an organic molecule may influence the activity of a site in a compound by hindering the attack of a reagent, e.g. 2,6-dimethyl benzoic acid has difficulty in forming an ester, as the methyl groups hinder the attack on the carboxyl groups, whereas 2,6-dimethyl phenylmethanoic acid readily forms an ester. Delocalization of π-orbitals can take place only if the orbitals can become parallel, or nearly so.

electrophilic attack an attack by an electrophilic reagent (p.205) on a substrate atom where electrons are readily available. Electrophilic attack is made by: H^+; H_3O^+; HNO_3; H_2SO_4; HNO_2; $AlCl_3$; $ZnCl_2$; $FeCl_3$; Br_2; I-Cl; CN-Cl, etc.; these reagents are electrophiles.

nucleophilic attack an attack by a nucleophilic reagent (p.205) on a substrate atom which lacks orbital electrons. Nucleophilic attack is made by: OH^-; CL^-; Br^-; HSO_3^-; CN^-; $R-C \equiv C^-$; $=O:$; $\equiv N:$, $=S:$, etc.

protonation (n) electrophilic attack (p.209) by a proton in which a proton is added to an electron-rich atom in an organic substrate (p.205), e.g. the intermediate stage in the preparation of bromoethane from ethanol.

organic reaction types reactions can be classified: (a) by the reagent used, e.g. hydrolysis, halogenation, methylation; (b) by the physical agent used, e.g. photochemical, pyrolysis; (c) by reaction mechanism. e.g. elimination, substitution, addition, rearrangement; (d) according to the product, e.g. polymerization. Further classification can be made (e) according to thermo-dynamic criteria, e.g. exothermic, endothermic; (f) according to kinetic considerations, e.g. zero order, first order, slow, fast.

substitution[2] (n) this generally refers to the substitution of an atom, or a group of atoms, in place of a hydrogen atom attached to a saturated carbon atom, but it can be in place of another atom or group of atoms. The substitution can be electrophilic, nucleophilic or radical-induced. *See* **substitution**[1] (p.208).

substituent (n) an atom, or group of atoms, that has been substituted in an organic compound, particularly an aromatic compound.

displacement (n) a substitution reaction (↑).

electrophilic substitution this is mainly the substitution of an atom, or a group of atoms, for hydrogen in a saturated, or aromatic, organic compound. Substitution in a benzene ring is a typical example.

nucleophilic substitution frequently this is the replacement of an atom other than hydrogen by another atom, or a group of atoms, e.g. the action of cyanides on alkyl halides in which the halogen atom is replaced by a cyano group: $C_2H_5Br + CN^- \rightarrow C_2H_5CN + Br^-$. Nucleophilic substitution of hydrogen is known in aromatic compounds. Radical-induced substitution is also known, e.g. the chlorination of alkanes.

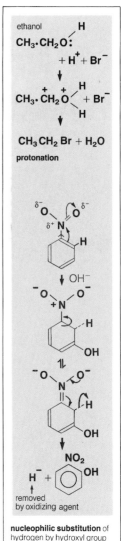

protonation

nucleophilic substitution of hydrogen by hydroxyl group

SE reaction an electrophilic substitution reaction (↑) using an electrophilic reagent. The nitration and sulphonation of benzene are typical examples.

SN1 reaction a substitution nucleophilic unimolecular reaction. A typical example is the conversion of a tertiary alkyl halide to an alcohol by the action of an alkali:

$RX + OH^- \rightarrow R - OH + X^-$. The rate of reaction is proportional to [RX] and is independent of [OH^-], so it is a unimolecular reaction. The mechanism is shown in the diagram. The more polar the solvent used in the reaction, the more likely is the substitution to be an SN1 reaction and not an SN2 reaction (↓). This is due to a polar solvent promoting ionization.

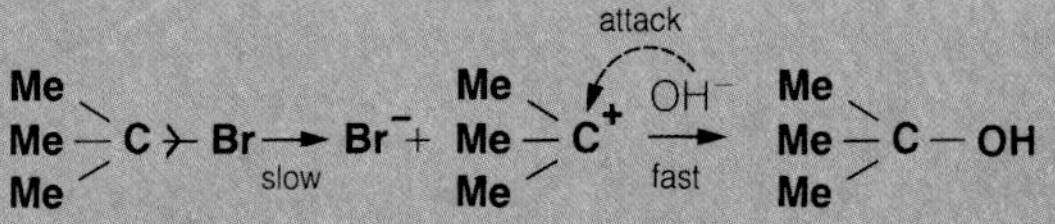

$SN1$ reaction

SN2 reaction a substitution nucleophilic bimolecular reaction. A typical example is the conversion of a primary alkyl halide to an alcohol by the action of alkali: $RX + OH^- \rightarrow R - OH + X^-$. The rate of reaction is proportional to [RX][OH^-], so it is a bimolecular reaction. The mechanism involves the formation of a transition state compound. The less polar a solvent used, the more likely is the substitution to be SN2 and not SN1 (↑).

$SN2$ reaction

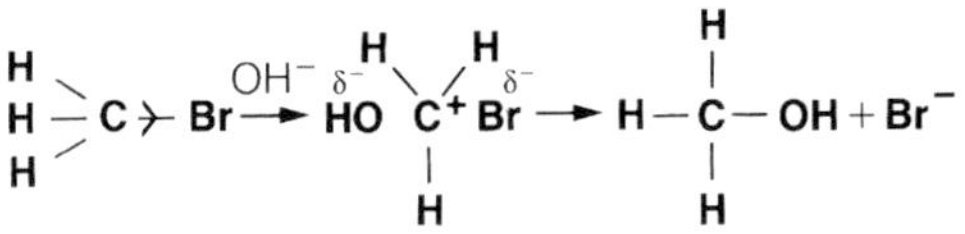

SNi reaction a substitution nucleophilic internal reaction. Both SN1 and SN2 reactions (↑) can produce inversion of configuration, or racemization (p.98) of the product. A rarer form of substitution, SNi forms products with the same configuration as the starting material. An example is the substitution of OH by Cl using sulphur dichloride oxide, $SOCl_2$.

addition (*n*) a chemical process in which two substances react to form a single substance. There are two main types of addition: (a) to a carbon-carbon double or triple bond, (b) to a carbon-oxygen double bond. The carbon-carbon double bond, also the triple bond, has π-bonds shielding the carbon atoms from nucleophilic attack, while the π-bonds are electron-rich and readily attacked by electrophilic reagents, e.g. acids. An ionizable compound $X^+ - Y^-$ attacks with its positive ion to form a π complex, *see diagram*, with a carbonium ion (p.215). The latter ion then combines with the negative ion of the reagent. If the unsaturated substrate (p.205) is asymmetrical, then the electron-donating alkyl groups influence the π complex so that the bond formed with the positive ion is pushed to the carbon atom with fewer alkyl groups, giving that carbon atom a negative character and leaving the other carbon atom with a positive character, which, in turn, attracts the negative ion of the reagent. Free radical (p.214) additions also occur with carbon-carbon double bonds. The carbon-oxygen double bond has two π-bonds, with a dipole moment, negative on the oxygen atom due to its electronegativity (p.75). Characteristic addition is initiated by nucleophilic attack on the positive carbon atom, e.g. by an anion such as CN^- or OH^-. The carbon atom becomes more positive if a proton combines with the oxygen atom, so the addition is catalysed by acids. Acidity, however, can reduce the ionization of the weak acids used to form adducts (↓), hence there is an optimum acidity, about pH = 4 for catalysing the reaction. The effect of alkyl groups (electron-donating) is to make the carbon atom less positive, i.e. less reactive, so ketones are less reactive than aldehydes. **add** (*v*).

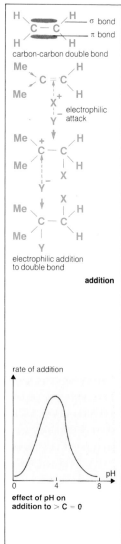

carbon-carbon double bond

electrophilic addition to double bond

addition

rate of addition

effect of pH on addition to $> C = O$

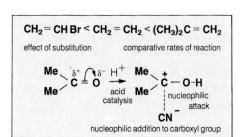

$$CH_2 = CH\,Br < CH_2 = CH_2 < (CH_3)_2 C = CH_2$$

effect of substitution   comparative rates of reaction

nucleophilic addition to carboxyl group

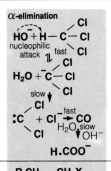

α-elimination

adduct (*n*) (1) the compound formed during addition (↑), e.g. when bromine undergoes an addition reaction with ethene, the product, dibromoethane, is an adduct. (2) the compound, a reagent, which is added to a substrate, e.g. bromine is an adduct to ethene.

elimination (*n*) a reaction in which two groups are removed from an organic molecule without being replaced by other groups. In most elimination reactions a proton and a nucleophile are removed resulting in the formation of a multiple bond. There are two categories, α-elimination and β-elimination.

α-**elimination** a much less common reaction than β-elimination (↓); in it two atoms are removed from the same carbon atom, e.g. the hydrolysis of trichloromethane.

β-**elimination** the most common type of elimination, so β is often omitted. The nucleophilic group is removed from the α-carbon atom and a proton from the β-carbon atom, e.g.

$$CH_3CH_2Br \xrightarrow[\text{alcohol}]{OH^-} CH_2 = CH_2$$

The reagent is a base dissolved in alcohol. Groups can be lost from atoms other than carbon. β-elimination can take place by E1 or E2 mechanisms (↓).

E1 reaction a unimolecular reaction similar to the S_N1 reaction (p.211); the rate of reaction is proportional to the concentration of the substrate, e.g. in the elimination of HX from RCH_2CH_2X, where X is a halogen, rate $\propto [RCH_2CH_2X]$ and independent of the concentration of the strong base used as a reagent. The first stage is the formation of a carbonium ion; this determines the reaction rate. The OH^- ions act as electron-pair donors towards hydrogen, removing the proton from the β-carbon to form a multiple bond. If the OH^- ions act as nucleophiles, a substitution reaction results. A polar solvent favours substitution, a non-polar solvent favours elimination.

E2 reaction a bimolecular reaction similar to the S_N2 reaction (p.211), e.g. the elimination of HX from $CH_3.CH_2X$, where X is a halogen, using a strong base. The rate of reaction $\propto [CH_3CH_2X][OH^-]$. The mechanism involves the abstraction of a proton from the β-carbon atom, accompanied by simultaneous loss of the halide ion from the α-carbon atom. The less polar a solvent, the more likely is the reaction to be E2 and not E1, compare **S_N1** and **S_N2 reactions** (p.211).

rearrangement (*n*) this process involves the migration of a functional group, or the rearrangement of the carbon skeleton of an organic compound, and it may proceed by means of electrophilic, nucleophilic, or radical intermediates. The actual rearrangement is frequently followed by a substitution, elimination or addition reaction to produce a stable end product. Many of the rearrangements consist of the breaking of a carbon-carbon bond, and the migration of the carbon group to form a new bond with an oxygen, nitrogen or another carbon atom. The migration is to an atom that is electron-deficient, and the general outline of a rearrangement is shown in the diagram. A simple example of rearrangement is:

$$CH_3.CH_2.CH = CH_2 \overset{H_2SO_4}{\rightleftharpoons} CH_3.CH = CH.CH_3$$

See **Beckmann transformation** (p.229), **enol-form** (p.33).

prototropy (*n*) a rearrangement (↑) which is an example of tautomerism (p.33) in which the mobile atom is a proton. Examples include the keto-enol rearrangement and the rearrangement of aliphatic nitro compounds.

electron-deficient species compounds containing atoms which do not have an electron octet in their valency shell. Boron in its compounds has a sextet; nitrogen in some compounds also has a sextet; the carbonium ion is also electron-deficient. Electron-deficient species are electrophiles (p.205).

heterolytic fission the breaking of a covalent bond so that one atom retains two electrons and the other atom retains none, e.g. it becomes an atom in an electron-deficient species (↑).

homolytic fission the breaking of a covalent bond so that each atom retains one electron, forming two free radicals (↓).

basic principle of **rearrangement**

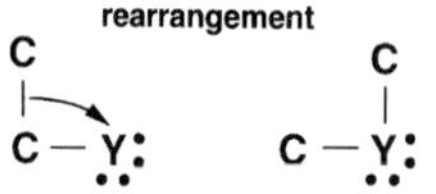

Y is an electron-deficient atom

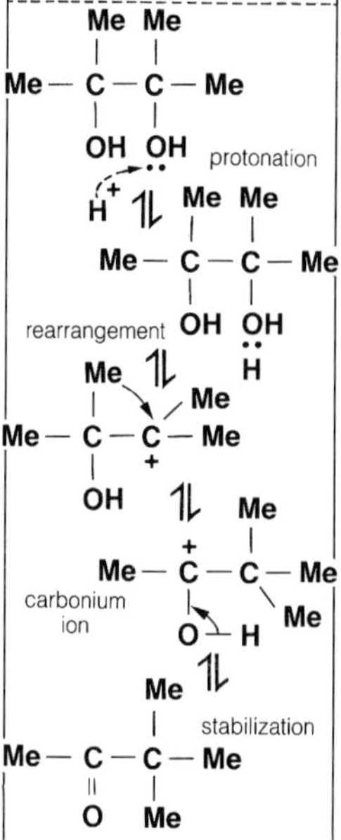

mechanism in rearrangement

type	fission	products		electro-negativity
heterolytic	$X \overset{..}{:} Y$	$X^- \overset{..}{:} + Y^+$	ions	$X > Y$
	$X \overset{..}{:} Y$	$X^+ + Y^- \overset{..}{:}$		$Y > X$
homolytic	$X \bullet \overset{|}{\bullet} Y$	$X \bullet + Y \bullet$	free radicals	$X \approx Y$

fission of organic molecules

free radical a radical formed by homolytic fission (↑); it possesses an unpaired electron and is highly reactive.

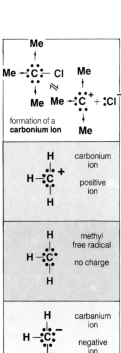

formation of a
carbonium ion

carbonium ion
positive ion

methyl free radical
no charge

carbanium ion
negative ion

formation of a
carbanion

The most stable odd-electron compounds are NO and NO$_2$. Organic free radicals have a life which ranges from comparatively long, e.g. triphenyl methyl radical, to very short, e.g. methyl radical with a life of 10^{-2} seconds. Free radicals generally have a transient life as important intermediaries in organic reactions.

homolysis (n) homolytic fission. **homolytic** (adj).

heterolysis (n) heterolytic fission. **heterolytic** (adj).

homolytic scission another name for **homolytic fission** (↑). It is a more accurate term as scission is equal splitting or equal fission.

thermolysis (n) the process of producing homolysis (↑) by heat. At convenient temperatures, i.e. < 150°C, homolysis occurs in relatively weak bonds, i.e. bonds of strength approximately 160 kJ mol^{-1}. Bonds of this strength are found in alkyl nitrites, nitrates (II), some azo compounds and peroxides. **thermolytic** (adj).

photolysis (n) the process of producing homolysis (↑) by electromagnetic radiation. Substances which can be photolysed include those which undergo thermolysis (↑), together with aryl iodides, organomercury derivatives RHgI and certain polyhalogenoalkanes, e.g C Cl$_3$Br. Carbonyl compounds absorb radiation at 320 nm and can be photo-excited by light at this wavelength.

carbonium ion the formation of an ion by heterolysis (↑) in which a carbon atom in an organic molecule loses the two electrons in a covalent bond and acquires a positive charge. Its existence is normally only transient, but the occurrence of such ions in organic reactions is widespread and of considerable importance in many chemical reactions. A carbonium ion is formed when a strongly electronegative atom, or group, is split off from a carbon atom.

carbocation (n) carbonium ion (↑)

carbanion (n) the formation of an ion by heterolysis (↑) in which a carbon atom in an organic molecule keeps the two electrons in a covalent bond and acquires a negative charge. If an organic molecule functions as an acid with a proton liberated from a carbon atom, a carbanion is formed and it is the conjugate base (p.203) of the proton. The presence of electron-withdrawing groups helps the formation of carbanions. As with carbonium ions (↑), the existence of carbanions is normally only transient, but they are important as intermediates in a wide variety of organic reactions.

radical reactions the reactions are characterized by the high rate of reaction, and the initiation and termination of the reaction by initiators and terminators respectively. Reactions are classified as unimolecular, bimolecular, radical-radical or radical chain reactions (↓). Unimolecular radical reactions occur with unstable radicals and generally decompose or undergo rearrangement before further action. Bimolecular reactions are SH2 reactions; radical-radical reactions are dimerization or radical disproportionation.

SH2 reaction substitution homolytic bimolecular reaction. SH2 reactions consist mostly of a homolytic attack on a univalent atom (usually hydrogen or a halogen) by a radical; they give rise to a new radical, e.g. the reaction between a phenyl radical and bromotrichloromethane: $Ph\cdot + CBrCl_3 \rightarrow \cdot CCl_3 + PhBr$

dimerization (n) the reaction between two identical radicals. The reaction between two different radicals is radical combination, or coupling. Dimerization of phenyl methyl radicals: $2PhCH_2\cdot \rightarrow PhCH_2CH_2Ph$ (1,2-diphenyl ethane); radical combination of trichloromethyl radicals with 1,1,1-trichloro-3-nonyl radicals:
$\cdot CCl_3 + C_6H_{13}CHCH_2CCl_3 \rightarrow C_6H_{13}CH(CCl_3)CH_2CCl_3$

radical chain reaction a substance, an initiator (↓), starts a chain reaction. A propagating reaction produces a cyclic process, repeated continuously, *see diagram*, until another substance, an inhibitor, stops the propagating reaction. Chains usually start with the abstraction of a hydrogen atom from an organic molecule, forming a free radical.

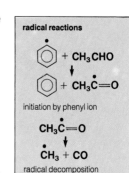

radical reactions

initiation by phenyl ion

radical decomposition

unimolecular radical decomposition

stage	reaction
initiation	$Cl - Cl \rightarrow 2Cl\bullet$ by photolysis
propagation	$Cl\bullet + CH_4 \rightarrow HCl + \overset{\bullet}{C}H_3$ $\overset{\bullet}{C}H_3 + Cl - Cl \rightarrow CH_3Cl + Cl\bullet$ $(Cl\bullet + CH_4$ repeats cycle $)$
termination	$2Cl\bullet \rightarrow Cl - Cl$ $\overset{\bullet}{C}H_3 + Cl\bullet \rightarrow CH_3Cl$ $\overset{\bullet}{C}H_3 + \overset{\bullet}{C}H_3 \rightarrow CH_3 - CH_3$ any one of these

radical chain reaction

initiator (*n*) a substance which readily undergoes homolytic scission (p.215) by thermolysis or photolysis, e.g. organic peroxides, azo compounds, organic nitrates (III) (nitrites). An initiator can act as a catalyst for a chain reaction. **initiate** (*v*).

inhibitor (*n*) a long-lived free radical or a substance that readily forms such radicals. An inhibitor reacts with short-lived radicals to terminate a chain reaction. Examples are: benzene-1,4-diol (hydroquinone), diphenylamine, iodine, 2,4,6-tri-tertiary butylphenol, and nitroxyls. **inhibit** (*v*).

oxonium ion the conjugate acid (p.203) of a base is the protonated base: the base possesses a lone pair, or lone pairs, of electrons which it can donate. In organic molecules, oxygen and nitrogen atoms are the usual sites of lone pairs. If a compound contains an oxygen atom it can act as a base and become protonated. Alcohols and ethers, and also water, can thus act as bases and may become protonated to form oxonium ions, e.g. R-O-R, where R is an alkyl group, or hydrogen, can form an oxonium ion R-OH-R, and oxonium salts. The hydronium, or hydroxonium, ion is a particular example of an oxonium ion. Oxonium ions are cations and may form salts with suitable anions.

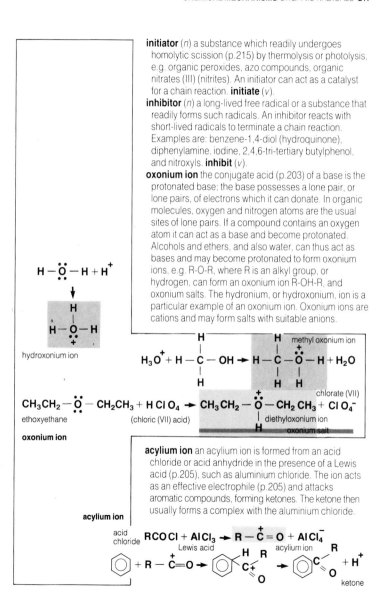

hydroxonium ion

ethoxyethane   (chloric (VII) acid)   diethyloxonium ion

oxonium ion

acylium ion an acylium ion is formed from an acid chloride or acid anhydride in the presence of a Lewis acid (p.205), such as aluminium chloride. The ion acts as an effective electrophile (p.205) and attacks aromatic compounds, forming ketones. The ketone then usually forms a complex with the aluminium chloride.

acylium ion

halonium ion a halogen ion with a positive charge. The compound Br-Cl ionizes thus: Br-Cl $\leftrightharpoons$ Br^+ + Cl^-, since chlorine is more electronegative than bromine. The positive bromine ion reacts with a benzene nucleus to substitute in the benzene ring. Similarly, the compound I-Cl substitutes iodine in a benzene ring. A Lewis acid (p.205), such as $ZnCl_2$, $AlBr_3$, $FeBr_3$, induces polarization in a halogen, and the positive end of the dipole attacks an aromatic compound.

bromonium ion an intermediate compound postulated in the addition of bromine to a double bond between carbon atoms. The π-electrons polarize the bromine molecule and the positive ion is an electrophile attacking the double bond. The formation of the bromonium ion is shown in the diagram. This favours the formation of a *trans*-compound, as the bulky bromine ion gives steric hindrance on the *cis*-side.

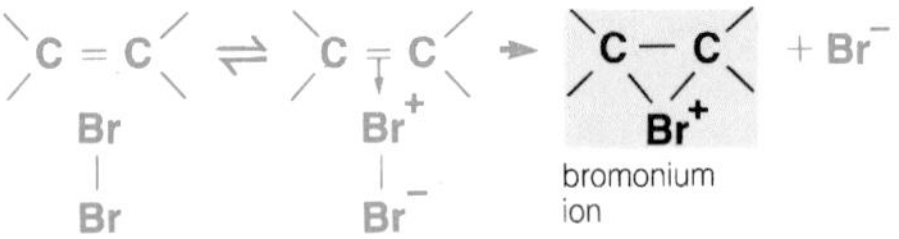

bromonium ion

nitronium ion an ion formed from concentrated nitric acid by the action of concentrated sulphuric acid, which protonates the nitric acid *see diagram*. The ion is a powerful electrophile and readily attacks a benzene ring; the ion *nitrates* the benzene ring.

$$HO-NO_2 + H_2SO_4 \rightarrow HSO_4^- + HO^+(H)-NO_2$$

nitric acid

$$\downarrow + H_2SO_4$$

$$HSO_4^- + H_3O^+ + NO_2^+$$

nitronium ion

nitronium ion

nitrosonium ion an ion formed from nitrous acid, HNO_2, by protonation; it is $\overset{+}{N}=O$ *see diagram*. The ion is an electrophile; it attacks a benzene nucleus and undergoes a nucleophilic reaction with nitrogen atoms in organic molecules. It *nitrosates* a compound.

nitrosate (*v*) *see* **nitrosonium ion** (↑).

bisulphonium ion an ion, $^+SO_3H$, formed from concentrated sulphuric acid, thought to be the electrophilic agent in sulphonation of a benzene ring. The evidence is more in favour of $^+SO_3$ being the agent.

nitrous acid

$$HO-NO + HY$$

$$\downarrow$$

$$HO^+(H)-NO + Y^-$$

$$\downarrow + HY$$

$$Y^- + H_3O + NO^+$$

nitrosonium ion

nitrosonium ion

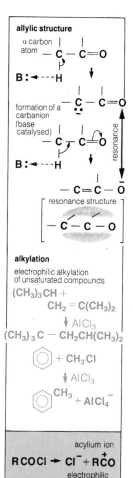

allylic structure

α carbon atom

formation of a carbanion (base catalysed)

resonance

resonance structure

alkylation

electrophilic alkylation of unsaturated compounds

$(CH_3)_3 CH +$
$CH_2 = C(CH_3)_2$

↓ $AlCl_3$

$(CH_3)_3 C - CH_2CH(CH_3)_2$

$+ CH_3 Cl$

↓ $AlCl_3$

CH_3 $+ AlCl_4^-$

acylium ion

$RCOCl \rightarrow Cl^- + R\overset{+}{C}O$

electrophilic attack

acetylation

allylic (*adj*) describes a structural system (or a reaction associated with that system) which has a saturated carbon atom adjacent to a carbonyl group (p.36), or to a double bond between carbon atoms. The saturated carbon atom is called an α-carbon. The carbonyl group has an electron-withdrawing capacity which is transmitted *see diagram*, to the α-carbon, giving any hydrogen atom attached to it a weak acidic nature. This electron shift delocalizes the π-electrons, which spread over the three-atom structure, and the carbanion (p.215) so formed is stabilized by resonance; the ion becomes an effective electrophile. Compounds with carbon-carbon double bonds can undergo allylic rearrangements, e.g. the carbonium ion formed from 3-chlorobut-1-ene has a resonance structure which undergoes solvolysis (p.142) with ethanol to form two ethers.

alkylation (*n*) the substitution of an alkyl group in place of hydrogen in an organic or inorganic molecule. The reagent used is an alkyl halide in the presence of a strong base such as sodamide. With compounds containing carboxyl groups the base forms a carbanion, and a nucleophilic substitution reaction (p.210) occurs; with other reactive groups, a carbonium ion is formed, and the α-carbon is the site of alkylation, e.g.

$$(CH_3)_2CHCN + CH_3I \xrightarrow{NaNH_2} (CH_3)_3C - CN + HI$$

With compounds containing carbon-carbon double bonds, and with aromatic compounds, an electrophilic substitution reaction takes place. $AlCl_3$ is used as a catalyst. *See* **Friedel-Crafts reaction**.

acetylation (*n*) the introduction of an acyl group (R.C = O) into a molecule containing −OH, −NH₂ or −SH groups. The acetylating agent used is ethanoyl chloride, CH_3COCl, or ethanoic anhydride for the substitution of the ethanoyl group, CH_3CO. The acylium ion (↑) formed from the acetylating agent attacks the α-carbon atom in allylic structures (↑) in an electrophilic reaction, e.g.

$$CH_3COCH_2COCH_3 + CH_3COCl \xrightarrow{NaOC_2H_5}$$
$$CH_3COC(COCH_3)HCOCH_3$$

The ion attacks the nitrogen atom in amines, *see diagram*, and the sulphur atom in thiols.

acylation (*n*) *see* **acetylation** (↑).

hydrogenation (*n*) implies the addition of gaseous hydrogen to alkenes and alkynes in the presence of a metallic catalyst, e.g. platinum, palladium, nickel. The addition reaction always forms a cis-compound. The metal adsorbs hydrogen and then adsorbs the alkene on its surface, so both hydrogen atoms approach on the same side of the double bond, *see diagram*. Other double bonds, such as C = O, C = N, can also be reduced catalytically.

catalytic hydrogenation hydrogenation (↑) with gaseous hydrogen.

dehydrogenation (*n*) implies the removal of hydrogen as a gas from an alkane, substituted alkane or an alcohol, in the presence of a catalyst such as: aluminium oxide, chromium oxide, zinc oxide, etc. The alkanes form alkenes and the alcohols form ketones.

catalytic dehydrogenation dehydrogenation (↑) producing gaseous hydrogen.

reduction (*n*) reducing agents are nucleophiles (p.205). Reduction by dissolving metals, e.g. sodium amalgam or sodium in ethanol, zinc dust in alkali, is brought about by the metal donating electrons to the substrate, and the subsequent neutralization of the negative charge by a proton donated by the solvent. Lithium aluminium hydride, $LiAlH_4$, acts as a carrier of hydride ions, H^-, and provides a nucleophilic attack on a carbonyl group in aldehydes, ketones and esters, *see diagram*. Hydrogenation (↑) is also used for reduction.

oxidation (*n*) oxidizing agents are electrophiles (p.205) and oxidation is catalysed by acids. Combustion leads to complete oxidation with all bonds broken with CO_2 and H_2O as the final products. The oxidation of an alcohol yields an aldehyde or a ketone. Mild oxidation adds −OH groups to both sides of a carbon-carbon double bond; this type of oxidation is also called **hydroxylation**.

hydroxylation (*n*) *see* **oxidation** (↑).

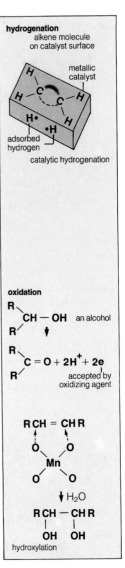

hydrogenation
alkene molecule on catalyst surface

catalytic hydrogenation

oxidation

an alcohol

accepted by oxidizing agent

hydroxylation

reduction

reduction of carbonyl group

reduction of an ester

saponification

$$OR'$$
$$R \overset{+}{-} C \overset{\frown}{=} O$$
$$HO:$$
$$\big\Updownarrow OH^-$$
$$OR'$$
$$R - C \overset{\frown}{-} \overset{..}{O}{}^-$$
$$OH$$
$$\big\Updownarrow$$
$$^-OR'$$
$$R \overset{+}{-} C = O$$
$$OH$$
$$\big\downarrow + OH^-$$
$$H_3O^+ + R - C = O$$
$$O^-$$

almost complete ionisation

$$R - C \overset{\frown}{=} O$$
$$\overset{|}{C}:NH_2 \quad \text{amide}$$
$$\big\Updownarrow$$
$$R - C - OH$$
$$\overset{||}{C}:NH \dashrightarrow :OH^-$$
$$\big\downarrow$$
$$R - C \overset{\frown}{-} OH$$
$$\overset{||}{C}:N$$
$$\big\downarrow H^+$$
$$H_2O + R - C$$
$$\overset{|||}{N} \quad \text{a nitrile}$$

dehydration
dehydration of an amide

ozonolysis (*n*) the addition of ozone to alkenes; it is an electrophilic addition reaction which is catalysed by Lewis acids (p.205). The reaction, produces an unstable ozonide, which is hydrolysed by boiling with water to produce two fragments. The carbonyl compounds, which are the end-products, are used in identifying the structure of an alkene.

esterification (*n*) the reaction between an organic acid and an alcohol to produce an ester; the reaction is catalysed by strong mineral acids. The mechanism consists of protonation (p.210) of the organic acid, which increases the ability of nucleophilic attack by the alcohol, which is then followed by dehydration and deprotonation of the intermediate compound. The reaction is reversible, e.g.
$CH_3CH_2COOH + ROH \rightleftharpoons CH_3CH_2COOR + H_2O$.
The determination of which product is obtained depends on the normal methods of moving an equilibrium point in a reversible reaction.

hydrolysis (*n*) the reverse process to esterification (↑). Hydrolysis can be alkali- or acid-catalysed; acid hydrolysis reaches an equilibrium point.

saponification (*n*) the alkaline hydrolysis (↑) of an ester. The hydroxyl ions of the alkali act as a nucleophile (p.205) and attack the carbonium ion formed from the ester. Further attack under alkaline conditions forms an ion of the acid, and ionization in the presence of a metal cation is practically complete.

dehydration (*n*) this is a nucleophilic E1 reaction (p.213) catalysed by acids for the dehydration of alcohols and by bases for the dehydration of amides (p.38). Phosphorus pentoxide and dialuminium trioxide are also used as dehydrating catalysts; they act as Lewis acids (p.205). The mechanism for alcohols starts with protonation, while the mechanism for amides involves proton loss to a base.

condensation (*n*) a reaction in which there is an addition reaction between two organic molecules followed by the elimination of a small molecule such as water or ammonia, or an alcohol. The addition reaction generally produces an unstable intermediate compound which loses water, e.g. the formation of oximes, phenylhydrazones, etc.:

$$CH_3CHO + H_2N.NHPh \rightarrow CH_3CH=N.NHPh.$$

See **Claisen condensation** (p.229).

nitration (*n*) nitration of the benzene ring is a substitution reaction; it is effected with a mixture of concentrated sulphuric acid and concentrated nitric acid. The reagent forms a nitronium ion (p.218), an electrophilic reagent; the ion attacks the concentration of negative charge formed by the delocalized π-orbitals of the benzene ring, *see diagram*. Highly reactive aromatic compounds, e.g. phenol, can be nitrated by dilute nitric acid due to the presence of the nitrosonium ion (p.218); the nitroso-compound formed is oxidised to the nitro-compound.

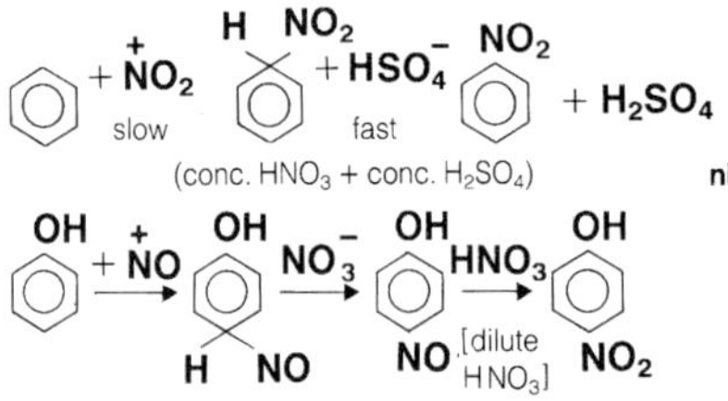

nitration

sulphonation (*n*) benzene is treated with fuming sulphuric acid; the reagent contains free sulphur (VI) oxide, a strong electrophilic reagent, which attacks the concentration of negative charge on the benzene ring, to form benzene sulphonic acid; the reaction is reversible.

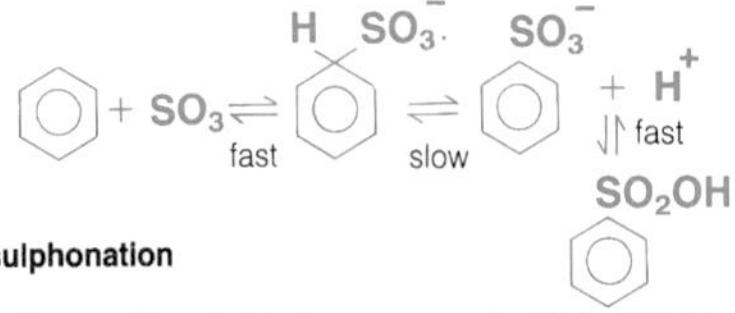

sulphonation

halogenation (*n*) halogens react with benzene only in the presence of a catalyst, which is usually a Lewis acid (p.205), e.g. $AlCl_3$, $ZnCl_2$, $FeBr_3$. The catalyst induces polarization in the halogen molecule, and the positive end attacks the π-electrons of the benzene ring as an electrophile (p.205). Confirmation of this mechanism is given by the compounds Br-Cl and I-Cl which brominate and iodinate, respectively, benzene; chlorine, being more electronegative than either bromine or iodine, causes polarization, with chlorine forming the negative end of each of the molecules. Halogenation can also be produced by aqueous

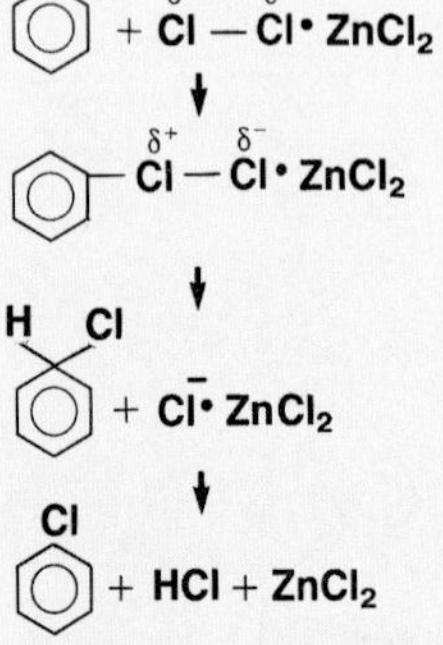

halogenation

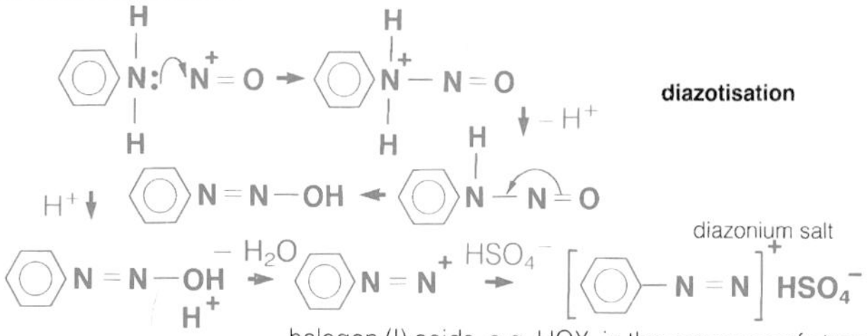

diazotisation

diazonium salt

halogen (I) acids, e.g. HOX, in the presence of strong mineral acids; ionization takes place as;

$$HO - X \xrightarrow{H^+} H_2O - X \rightarrow H_2O + X^+.$$

and the positive ion is the electrophilic reagent.

diazotisation (*n*) the production of a diazonium ion ($R - N_2^+$) from a primary amine by the electrophilic attack of a nitrosonium ion (p.218). Aliphatic diazonium ions are highly unstable, due to the great stability of N_2, but aromatic diazonium ions are relatively stable because of delocalization of charge through the π-orbital electron system of the aromatic ring. The nitrosonium ion is formed from sodium nitrite, $NaNO_2$, and excess strong mineral acid. The reaction is carried out at a low temperature to inhibit side reactions from the highly reactive aromatic diazonium ion and the product is a diazonium salt, which usually is made to undergo a further reaction.

diazo coupling diazonium salts couple with phenols in the 4-position, or the 2,6-position if the 4-position is occupied, to form azo compounds. The salts couple with primary and secondary aromatic amines to form diazo-amino compounds. The diazonium ion is a weak electrophile (p.205) and it attacks only highly reactive aromatic compounds such as phenols and amines. Electron-withdrawing groups (p.209) substituted at 2,6-, or 4-positions increase the electrophilic nature of the diazonium ion. Amines are less nucleophilic than phenols, and the nitrogen atom of the amine, and not a carbon atom, is attacked by the electrophilic ion. The pH of the reactants is important. With phenols, a slightly alkaline solution is used; a high pH forms a phenate, a low pH reduces the concentration of the diazonium ion. With amines, a slightly acid solution is used; a high pH forms PhN:NOH, which is not electrophilic, a low pH forms $PhNH_3^+$.

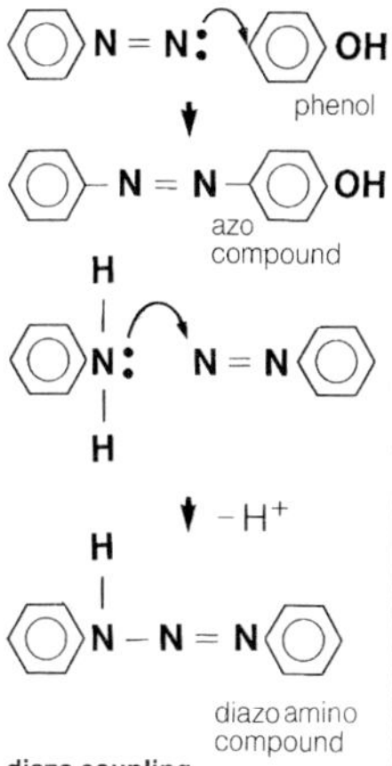

phenol

azo compound

diazo amino compound

diazo coupling

activating groups substituent groups which increase the reactivity of a benzene ring. The ring activity is due to the π-orbital electrons, so groups supplying electrons activate it. The inductive effect (p.75) of electron-donating groups activates the ring. The mesomeric effect (p.77) takes place if the atom attached to the benzene ring has an unshared pair of electrons; these interact with the delocalized π-electrons and increase electron-availability. The overall effect depends on both inductive and mesomeric effects and cannot be forecast; the dipole moment of a substituted compound gives an indication. For the NH_2 group, the two effects are in opposition, but the overall effect is electron-donating, as nitrogen readily releases its electron pairs; this is indicated by the dipole moment.

deactivating groups substituents which decrease the reactivity of a benzene ring. The inductive effect of electron-withdrawing groups deactivates the ring. The mesomeric effect deactivates the ring if the atom attached to the ring is itself multiple-bonded to a more electronegative atom. For the NO_2 group, the two effects reinforce each other, as shown by the dipole moment.

orienting effect a substituent already present in a benzene ring has an orienting effect on a further substitution. In general, the inductive effect (p.75) of the substituent controls the rate of substitution, but the mesomeric effect (p.77) controls the stabilization of the π-electrons, and governs the orientation of substitution. The mesomeric effect of deactivating groups (p.224) leads to 3-position substitution, and of activating groups to 2,4-position substitution. A further description of the mechanism considers the intermediate compounds formed by addition, and which of their compounds stabilize the delocalized π-electrons. All substitutions are relative in their attack on the benzene ring, and although one set of substitutions is preponderant, rarely is it exclusive. Steric effects can also affect substitution by hindrance in the 2-, or 6-position.

ortho-position the 2-, or 6-position of substitution.
para-position the 4-position of substitution.
meta-position the 3-, or 5-position of substitution.
mechanism (*n*) an explanation of the chemical processes by which a reaction proceeds, based on

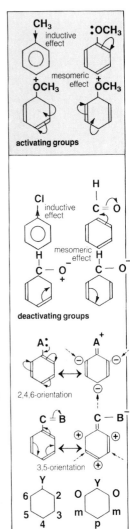

activating groups

deactivating groups

2,4-orientation

3,5-orientation

orientation effect

describing points of electron excess or deficiency in a molecule. The electron displacements in a molecule result from the inductive (p.75), mesomeric (p.77), inductomeric (p.208), and electromeric (p.209) effects. A slight change in electron density can alter the course of a reaction. The real test of a mechanism lies in its ability to forecast the rate and course of a reaction.

isomerisation (*n*) a change from one isomer to another. Structural isomerisation of the alkanes takes place in the presence of aluminium chloride with straight chains being converted to branched chains. Geometric isomerisation takes place in unsaturated compounds.

isomerization

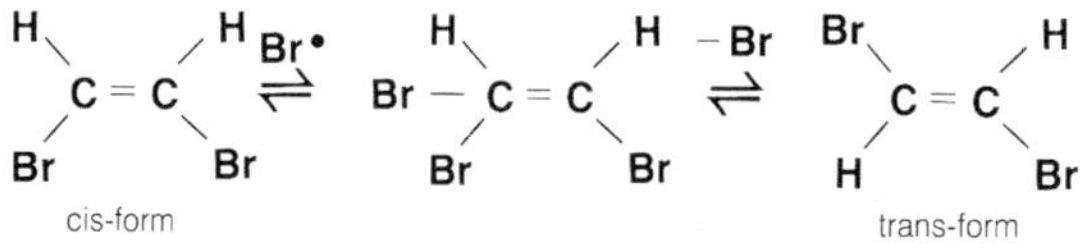

cis-form   trans-form

Markownikov's rules the first rule deals with the addition of halogen acids to alkenes, and the second rule with the addition of halogen acids to halogen-substituted alkenes. The empirical generalization is: in the addition of unsymmetrical adducts to unsymmetrical alkenes, the more electronegative group, including a halogen, becomes attached to the carbon atom carrying the smaller number of hydrogen atoms, or to the carbon atom combined with a halogen atom. A secondary carbonium ion (p.215) is formed in the reaction and the halogen, or a negative ion, makes a nucleophilic attack on it. This mechanism produces both products of Markownikov's rules.

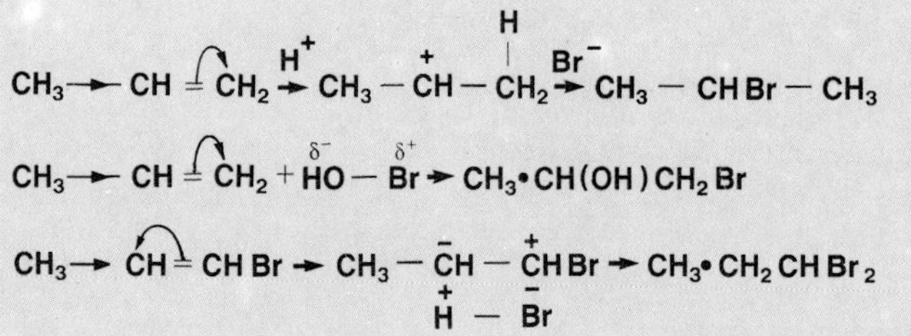

Markownikov's rules

dehalogenation (*n*) the action of zinc in warm alcohol on 1,2-dihalides, particularly bromides, forms alkenes. The zinc ion donates electrons to a bromine atom, and removes it from the molecule.

Würtz reaction the action of metallic sodium on an alkyl halide dissolved in an inert solvent, e.g. petrol. An alkyl sodium is formed as an intermediate, and then an alkane:

$C_2H_5Cl + 2Na \rightarrow C_2H_5Na + NaCl$
$C_2H_5Na + C_2H_5Cl \rightarrow C_2H_5.C_2H_5 + NaCl$

The product is contaminated with alkenes; tertiary halides do not provide a useful reaction. The mechanism of the reaction is ionic; using two alkyl halides, a mixture of alkanes can be formed, e.g. R-R, R-R', R'-R'.

Würtz-Fittig reaction the action of metallic sodium on equimolecular proportions of an aliphatic halide and an aromatic halide dissolved in ether (ethoxyethane). Either bromides or iodides are used; the reaction is slow, taking a few days, but it gives satisfactory results. The product is an alkyl benzene, with by-products of alkanes and aromatic hydrocarbons. The mechanism is ionic, as for the Würtz reaction. Two aromatic halides cannot be coupled by the reaction.

Fittig reaction *see* **Würtz-Fittig reaction** (↑).

Kolbé's electrolytic reaction the electrolysis of a concentrated aqueous solution of the sodium or potassium salt of an aliphatic carboxylic acid produces alkanes. Platinum electrodes are used. The reaction proceeds by a free radical mechanism (p.214). There is no reaction with aromatic carboxylic acids.

haloform reaction the reaction of a ketone, containing the grouping $-COCH_3$, with an alkaline solution of a halogen to form a trihalogenomethane, CHX_3, and a salt of a carboxylic acid. The halogen in alkali forms a salt such as NaClO, NaBrO or NaIO. A reaction for chlorine is:

$C_2H_5COCH_3 + 3NaClO \rightarrow C_2H_5COO^- + Na^+ + CHCl_3$.

The mechanism follows the formation of a carbanion (p.215) and further base-induced halogenation (p.222).

Williamson's synthesis the reaction between a sodium alkoxide and an alkyl halide, preferably an iodoalkane, to produce an ether, e.g.

$CH_3I + C_2H_5ONa \rightarrow CH_3OC_2H_5$.

Both mixed and simple ethers can be formed by this method. Mixed aliphatic and aromatic ethers can also

dehalogenation

be prepared: $PhONa + C_2H_5I \rightarrow PhOC_2H_5 + NaI$.
Aromatic ethers cannot be prepared in this way. The
alkali metal alkoxides are strong bases and strong
nucleophiles; they undergo nucleophilic substitution
with primary alkyl halides; it is an S_N1 type reaction.

$$CH_3.CH_2 - O - H + H^{+\cdot} \rightleftharpoons CH_3.CH_2 \overset{+}{\underset{\overset{\cdot\cdot}{H}}{O}} H \rightleftharpoons \overset{+}{C}H_3.CH_2 + H_2O$$

Williamson's continuous process

$$CH_3 \overset{+}{C}H_2 + HO - C_2H_5 \rightleftharpoons C_2H_5 - \overset{+}{\underset{\overset{\cdot\cdot}{H}}{O}} - C_2H_5 \xrightarrow{-H^+} C_2H_5OC_2H_5$$

Williamson's continuous process the action of
concentrated sulphuric acid on excess of an alcohol
produces a simple ether. The method is suitable for
lower primary alcohols only, as higher alcohols yield
alkenes. The alcohol is run continually into a mixture of
acid and alcohol, and produces a continuous supply of
ether. Protonation of the alcohol leads to the formation
of a carbonium ion (p.215) which is attacked by the
nucleophilic alcohol; the process is acid-catalysed.

Hofmann reaction the action of chlorine or bromine, in
the presence of hot aqueous alkali, on aliphatic and
aromatic amides to produce amines. The halogen and
the alkali form HClO or HBrO in solution. If cold
aqueous alkali is used, a halogen substitution is made
in the amino group. The mechanism for the reaction
consists of the formation of a halogen-substituted
amino group, then an isocyanate; this is followed by
hydration and decarboxylation (p.228). The isocyanate
can be isolated under suitable conditions.

Hofmann reaction

Hofmann degradation a quaternary ammonium
hydroxide solid salt is heated to 100°C, or an aqueous
solution of the salt is boiled. The solution is prepared by
treating a quaternary ammonium iodide with a
suspension of silver oxide in water or alcohol. An
elimination reaction (p.213) takes place and the
products are an alkene and a tertiary amine. This
reaction is useful for the determination of the structure
of nitrogen-containing compounds, particularly
alkaloids. This use is indicated in the diagram.

$$\left[(CH_3)_3 \overset{+}{N} C_3H_7\right] OH^- \xrightarrow{100°C} N(CH_3)_3 + C_3H_6 + H_2O$$

quaternary ammonium hydroxide   amine   propane

Hofmann degradation

$$R{\cdot}CH_2 - CH - CH_3 \rightarrow R{\cdot}CH_2\,CH = CH_2 + N(C_2H_5)_3$$
$$|$$
$$N(C_2H_5)_3$$

decarboxylation (n) the removal of the carboxyl group,
−COOH, from a carboxylic acid via the anions. The
mechanism involves the carbanion intermediate
acquiring a proton from the solvent.
Aromatic carboxylic acids and dicarboxylic acids are
decarboxylated by heat alone.

Beckmann rearrangement the conversion of a ketoxime
to an amide by the migration of an alkyl group from
carbon to nitrogen, catalysed by a wide variety of acids,
including Lewis acids, e.g. H_2SO_4, P_2O_5, BF_3, PCl_5. It is
not the electron-donating ability of the alkyl group that
determines the rearrangement, but the stereochemical
arrangement, as the *anti*-alkyl group makes a
trans-migration. The rearrangement starts with

Beckmann rearrangement

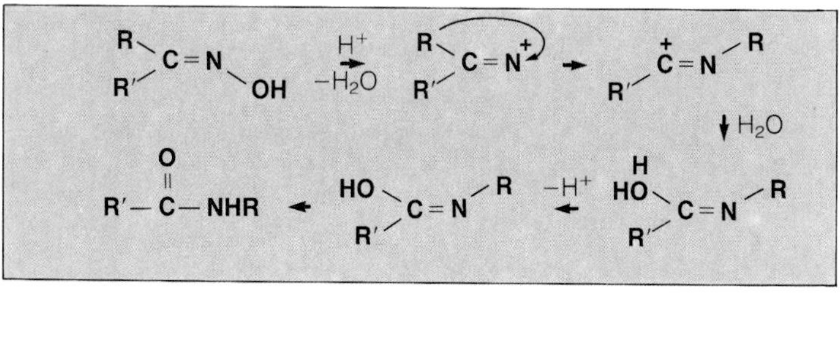

protonation and then dehydration of the ketoxime, followed by an intramolecular rearrangement and completed by the addition of water, deprotonation, and rearrangement of a hydrogen atom.

Beckmann transformation *see* **Beckmann rearrangement** (↑).

aldol reaction ethanal, CH_3CHO, on treatment with a trace of potassium hydroxide undergoes an intermolecular addition reaction producing 3-hydroxylbutanal (aldol). The reaction is: $CH_3.CHO + CH_3.CHO \rightleftharpoons CH_3.CH(OH).CH_2.CHO$. Propanone undergoes a similar reaction. These are aldol reactions, and they are not confined to intermolecular addition as two different molecules, even an aldehyde and a ketone, can undergo an aldol reaction. The mechanism consists of an addition to a carbon-oxygen double bond, through a carbanion attacking a carbonyl carbon atom. An aldol reaction is reversible.

aldol reaction

aldol condensation *see* **aldol reaction** (↑).

Claisen condensation ethyl ethanoate in the presence of sodium ethoxide undergoes an intermolecular condensation reaction to produce ethyl 3-oxobutanoate (ethyl acetoacetate). The reaction is:

$CH_3.COOC_2H_5 + CH_3.COOC_2H_5 \rightleftharpoons$
$CH_3COCH_2COOC_2H_5 + C_2H_5OH$

An ester group activates the α-carbon atom and its attached hydrogen atoms in a molecule. The effect is weak and needs reinforcing by other activating functional groups. The condensation is an aldol (↑) type reaction with a similar mechanism. Crossed condensation of two different esters can be carried out, but results in mixed products unless one of the esters is incapable of forming a carbanion. Claisen condensations are reversible.

Claisen ester condensation *see* **Claisen condensation** (↑).

Diels-Alder reaction this is a reaction carried out in a polar solvent between a 1,3-diene (an alkene with two double bonds) which has electron-donating substituents, and a compound called a dienophile. Typical dienophiles have one double bond, and need electron-withdrawing group (p.209) substituents to promote the reaction. The products are cyclic or aromatic compounds. Steric effects from the diene may influence the reaction. For bulky substituents only *trans/trans* configurations react; *cis/trans* and *cis/cis* configurations provide steric hindrance.

diene synthesis *see* **Diels-Alder reaction** (↑).

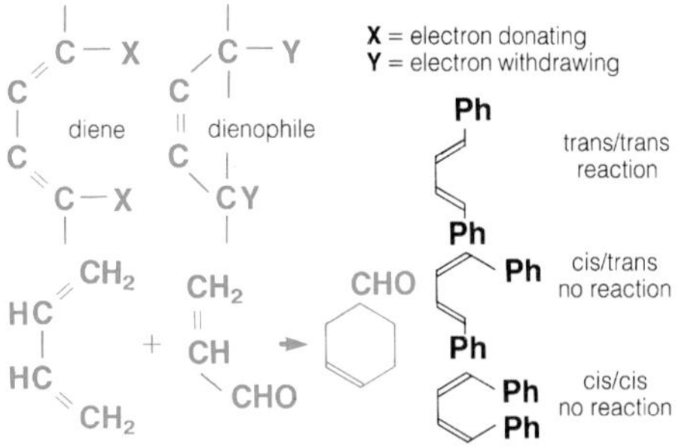

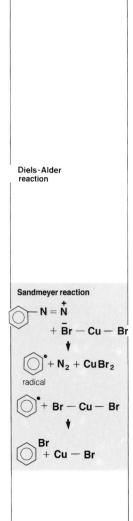

Diels-Alder reaction

Sandmeyer reaction

Sandmeyer reaction the replacement of the diazonium group (p.41) by a halogen or a cyano group. A diazonium salt in aqueous solution is warmed with a solution of a copper (I) halide dissolved in the corresponding halogen acid, e.g. Cu_2Br_2 in HBr. The copper salt is not necessary for replacement by iodine. For the cyano group, copper (I) cyanide is dissolved in a solution of sodium cyanide. The mechanism involves the formation of an aryl radical, followed by a displacement reaction. No chain reaction is promoted as the radical that is consumed does not lead to a second radical being formed.

Gattermann reaction a diazonium salt, the chloride or the bromide, is warmed with copper powder, and the chloro- or bromo-substituted benzene is formed:

$$PhN_2^+Cl^- \rightarrow Ph-Cl + N_2.$$

Yields are not as good as those for the Sandmeyer reaction (↑).

Friedel-Crafts reaction a reaction for the alkylation or acylation of benzene, substituted benzene compounds and benzene homologues. In alkylation, an aromatic hydrocarbon in the presence of aluminium chloride (used as a catalyst) reacts with an alkyl halide to form an alkyl-substituted aromatic hydrocarbon, e.g. $PhH + C_2H_5Cl \rightarrow PhC_2H_5 + HCl$. The reaction has the following drawbacks: (1) polyalkylation may take place as the initial substitution increases the reactivity of a benzene ring; (2) isomerization may take place, as alkane chains tend to undergo rearrangement to form branched isenes; (3) migration of alkyl groups is promoted by aluminium chloride, so different substitution may be effected instead of the required substitution. Effect (1) can be reduced by excess hydrocarbon. Effects (2) and (3) can be minimized by low temperatures. The mechanism of the reaction resembles halogenation (p.222) of a benzene ring. In acylation, an alkyl chloride or an acid anhydride, in the presence of a Lewis acid (p.205) forms an acylium ion (p.217) which makes an electrophilic attack (p.209) on the benzene ring. A ketone is formed which complexes with the aluminium chloride, and prevents complexing with the acylium ion. Alkylation can be effected by unsaturated hydrocarbons in the presence of a Lewis acid; BF_3 is generally used. The acid protonates the unsaturated hydrocarbon forming a carbonium ion (p.215) which makes an electrophilic attack on the benzene ring.

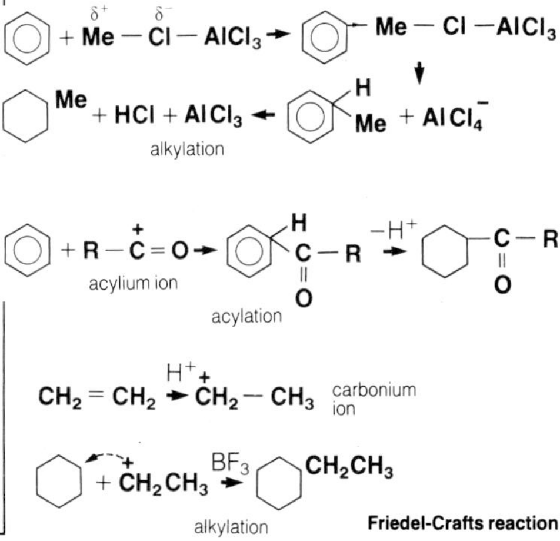

alkylation

acylium ion

acylation

$$CH_2 = CH_2 \xrightarrow{H^+} \overset{+}{C}H_2 - CH_3 \quad \text{carbonium ion}$$

alkylation

Friedel-Crafts reaction

Clemmensen reduction the reduction of the carbonyl group of aldehydes and ketones by the action of amalgamated zinc and hydrochloric acid. The substrate is reduced to a hydrocarbon. The reduction is mainly used for aromatic aldehydes and ketones, but can be used for aliphatic compounds. The reduction is the preferred method for the preparation of alkyl benzenes and related compounds, instead of the Friedel-Crafts reaction (p.230). Phenylethanone is reduced to ethyl benzene: $Ph.CO.CH_3 + 4H \rightarrow Ph.CH_2.CH_3$.

Cannizzaro reaction the reaction between aldehydes, with no hydrogen atom on the α-carbon atom in a molecule, and cold concentrated aqueous or alcoholic sodium hydroxide. A disproportionation of two molecules occurs, one molecule is reduced to an alcohol and the other molecule is oxidized to the salt of a carboxylic acid, e.g.

$2R_3C.CHO + NaOH \rightarrow R_3C.CH_2OH + R_3C.COONa$.

protective group a reactive functional group may take part in a reaction when another group, or structural feature, is the objective of the reaction. In this case, the reactive functional group is protected by a prior reaction in which a protective group is added to the functional group. A halogen atom in a molecule is protected by the action of sodium methoxide to form a methoxyether. A carbonyl group is protected by converting the compound into an acetal by the action of ethanol, or other alcohol. An amino group is protected by acylation, through the action of ethanoyl chloride. A double bond is protected by bromination, followed by

$$\textbf{RX} \xrightarrow{\text{Na OCH}_3} \textbf{R·O·CH}_3 \xrightarrow{\text{Hx}} \textbf{RX}$$

$$\textbf{R·CHO} \xrightarrow{\text{EtOH}} \textbf{R·CH(OEt)}_2 \xrightarrow[\text{aq}]{\text{HCl}} \textbf{R·CHO}$$

$$\textbf{R·NH}_2 \xrightarrow{\text{CH}_3\text{COCl}} \textbf{R·CONHCH}_3 \xrightarrow{\text{HCl}} \textbf{R·NH}_2$$

protective groups

debromination after a required action has been completed. A position on a benzene ring is protected by sulphonation; after the required action is completed, the compound is desulphonated.

Grignard reagent a solution of an alkyl halide, or an aryl bromide or iodide in ether is slowly run into a suspension of dry magnesium turnings in ether. A reaction occurs spontaneously with the evolution of heat, and a Grignard compound, RMgX, is formed, where X is a halogen. The compound is not isolated; the next reagent is added to produce a product. The mechanism for synthetic applications suggests that the Grignard compound is polarized with a positive magnesium and a negative carbon atom in the organic group. Addition reactions occur with carbonyl groups, or with cyano groups.

$$R-X + Mg \rightarrow \overset{\delta^-}{R}-\overset{\delta'}{Mg}-X \quad \text{Grignard reagent}$$

an alcohol

$$\text{Grignard reagent} \quad \overset{\delta^+}{\underset{}{>}}C=\overset{\delta^-}{O} + \overset{\delta^-}{R}-\overset{\delta^+}{MgX} \rightarrow \overset{R}{\underset{OMgX}{>C<}} \rightarrow \overset{R}{\underset{OH}{>C<}}$$

$$C\overset{O}{\underset{O}{<}} + \overset{\delta^-}{R}-\overset{\delta^+}{MgX} \rightarrow R-C\overset{O}{\underset{MgX}{<}} \rightarrow R-C\overset{O}{\underset{OH}{<}}$$

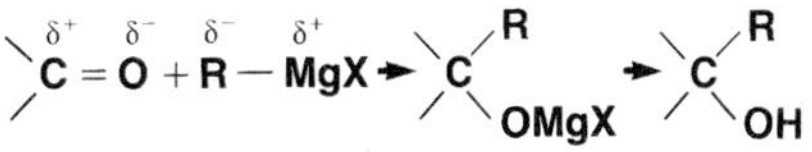

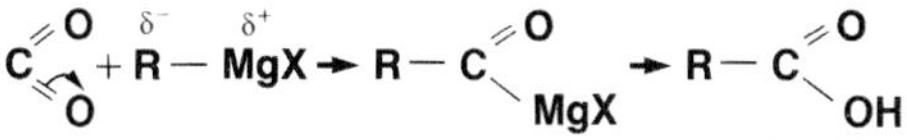

carboxylic acid

Étard's reagent chromium (VI) dichloride dioxide, CrO_2Cl_2. It is used to produce aromatic aldehydes directly from alkylated homologues of benzene, e.g. methyl benzene is converted to benzaldehyde. An intermediate chromium complex is formed, which is treated with water to hydrolyze it.

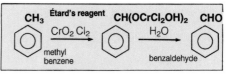

Schiff's reagent a solution of fuchsin (a basic triphenylamine dye) which has been decolourized by sulphur dioxide. Aliphatic aldehydes and aldose sugars restore the colour; aromatic aldehydes and aliphatic ketones restore the colour more slowly; aromatic ketones have no action.

carbylamine test a test for aliphatic and aromatic primary amines. The amine is warmed with trichloromethane and sodium hydroxide solution. An isocyano-compound is formed; it has a nauseating odour characteristic of all isocyano-compounds.

phenylamine → isocyanobenzene

carbylamine test

Lassaigne test an organic compound is fused with molten sodium. Any nitrogen in the compound forms sodium cyanide, and sodium hydroxide is also formed. Iron (II) sulphate is added to the mixture and, initially, sodium hexacyanoferrate (II) is formed which, on boiling, combines with Fe^{3+} ions formed to give a dark blue precipitate. Any sulphur in the compound forms sodium sulphide; the mixture is acidified with ethanoic acid and boiled, liberating hydrogen sulphide which is tested by lead (II) ethanoate paper. Any halogen in the compound is converted to a sodium halide. A solution of the mixture is tested with silver nitrate solution and dilute nitric acid. A white or yellow precipitate indicates a halogen present. If sulphur or nitrogen, or both, have been identified, the solution must be acidified and boiled to remove any cyanide or sulphide.

Fehling's test Fehling's solution is prepared by mixing a solution of copper (II) sulphate with an aqueous solution of sodium hydroxide and sodium potassium dihydroxybutanedioate (tartrate); it is copper (II) hydroxide forming a stable complex with the organic salt in an alkaline solution. A compound which is readily oxidized reduces the copper (II) ion to a copper (I) ion forming a red precipitate of copper (I) oxide. The test detects organic reducing agents such as aliphatic aldehydes, and compounds containing the structure −COCHOH. Simple ketones do not react to the test; neither do aromatic aldehydes. Monosaccharides and reducing disaccharides give a positive reaction with Fehling's solution.

Benedict's test Benedict's solution is an aqueous solution of copper (II) sulphate, sodium carbonate and sodium 2-hydroxypropane-1,2,3-tricarboxylate (citrate). The solution is stable; it is copper (II) hydroxide forming a stable complex with the organic salt in an alkaline solution. A compound which is readily oxidized reduces the copper (II) ion to a copper (I) ion forming a red precipitate of copper (I) oxide. The test is identical with Fehling's test (↑).

Tollen's reagent the reagent is prepared by adding ammonia solution to a solution of silver nitrate until the precipitate of silver oxide just dissolves. The silver is present in solution as a complex ion $[Ag(NH_3)_2]^+$. The reagent is readily reduced by easily oxidized organic compounds to form metallic silver, usually precipitated as a silver mirror on a glass container. The test is identical with Fehling's test (↑) for testing alkyl compounds and sugars.

Schweitzer's reagent ammonia solution is added to a solution of copper (II) sulphate until the precipitate of copper (II) hydroxide just dissolves, forming a complex ion tetraammine-copper (II), $[Cu(NH_3)_4]^{2+}$. Cellulose dissolves in the reagent. When sulphuric acid is added to this solution, cellulose is precipitated. The reagent is used in processes for making artificial fibres.

biuret test sodium hydroxide solution is added to the substance to be tested followed by one or two drops of 1% copper (II) sulphate solution. A substance containing two −CO–NH– groups attached to one another, or to the same nitrogen atom or the same carbon atom, gives a pink or violet colour for a positive reaction. The solutions test for biuret, amino acids, polypeptides and proteins.

iodine test starch consists of two structurally dissimilar polysaccharides (p.42) called amylose and amylopectin; the amylose content is 15–25%. Iodine with amylose produces an intense blue colour, and hence is used as a test for starch. With amylopectin, iodine produces a pale red colour. Iodine can be used for the quantitative analysis of various starches to determine the percentage of amylose.

photochemical reaction a chemical reaction that takes place when the system is exposed to light, i.e. taken as radiation between the wavelengths of 100–1000nm.

photochemical decomposition the decomposition of a compound by a photochemical reaction. Also called **photolysis** (p.215).

photochemical dissociation photochemical decomposition (↑), particularly for the decomposition of molecules into atoms, e.g. $Cl_2 \rightarrow 2Cl\cdot$ (free radical).

biuret test

polymerization (*n*) the formation of long chains and networks (p.110) to give molecules of high relative molecular mass (p.29) by combining together simple molecules by addition (p.212) and condensation (p.221) reactions in a continuous series of repeat units.

monomer (*n*) refers to a simple molecule with a low relative molecular mass (p.29) which can undergo polymerization (↑) reactions, e.g. styrene is the monomer for the polymer (↓) polystyrene, and ethene is the monomer for polyethylene (polythene). **monomeric** (*adj*).

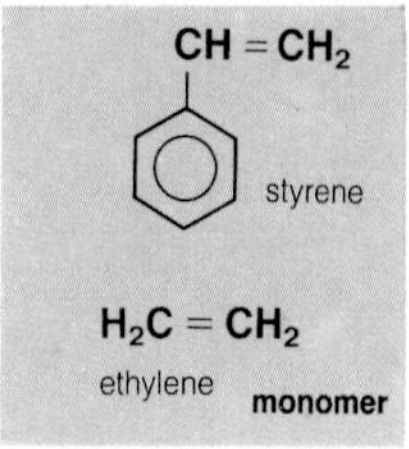

dimer (*n*) a molecule or compound resulting from the chemical combination of two molecules of the same substance, e.g. aluminium trichloride, $AlCl_3$, exists in the dimeric form as Al_2Cl_6 at low temperatures. **dimerization** (*n*), **dimerize** (*v*).

polymer (*n*) a substance of high relative molecular mass (p.29) formed from a series of repeating simple molecular units by a process of polymerization (↑). The polymer may be a natural one, e.g. cellulose, proteins, rubber; or semi-synthetic, e.g. nitrocellulose; or synthetic, e.g. polythene, nylon, polyvinyl chloride.

mer (*n*) refers to the repeating structural unit or units which go to make a polymer (↑), e.g. in polyethene the repeating unit is $-(CH_2-CH_2)-$, whilst nylon has two repeating units $-(NH(CH_2)_6NH)-$ and $-(CO(CH_2)_5CO)-$.

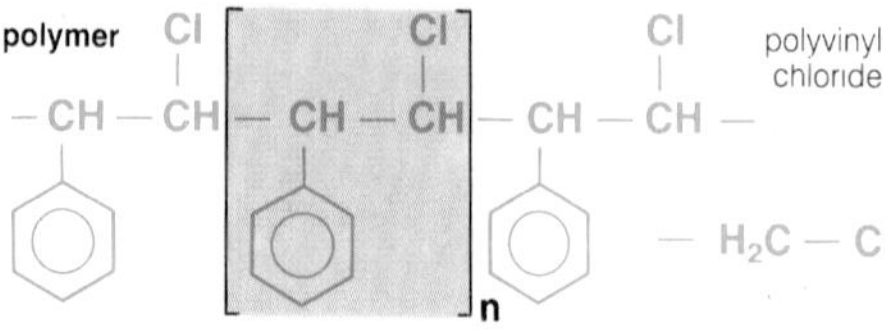

styrene

ethylene   **monomer**

polymer   polyvinyl chloride

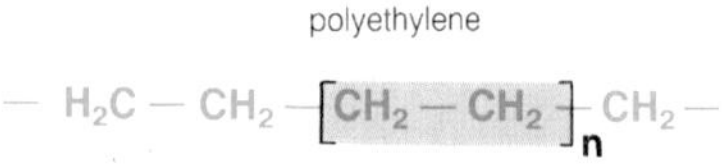

polyethylene

initiation stage the first step in a polymerization (↑) chain reaction (p.206) in which an initiator (↓) provides free radicals (p.214) which break the double bonds in unsaturated molecules to form alkene radicals which can then combine with other alkene (p.34) molecules to start the polymerization process.

initiation rate refers to the rate of formation of free radicals (p.214) from an initiator (↓) which can start a polymerization (↑) reaction.

free radical   **initiation stage**

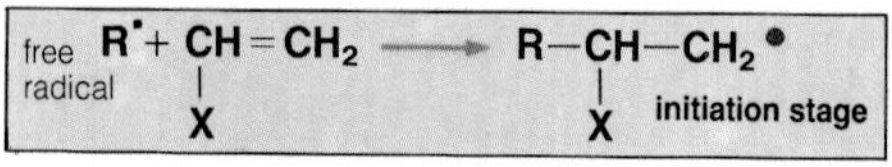

initiator (*n*) a substance which will readily form free radicals (p.214) when added to a monomer (↑) and serve to initiate a polymerization (↑) reaction. Most initiators are peroxides, e.g benzoyl peroxide is used in the polymerization of alkenes (p.34).

photosensitizer (*n*) a substance which is excited by light and able to transfer its excitation energy to polymer molecules, or to nearby oxygen atoms, thus promoting degradation of the polymer directly, or through oxidation (p.220) of the polymer (↑).

propagation stage the second stage in a polymerization chain reaction (p.206) in which the alkene (p.34) radical, formed by the action of the initiator (↑), combines with another unsaturated molecule to form a further radical which can then continue the process until all the monomer (↑) is exhausted, or the chain is terminated. **propagate** (*v*).

propagation stage

$$R-\underset{X}{CH}-CH_2^\bullet + \underset{X}{CH}=CH_2 \longrightarrow R-\underset{X}{CH}-CH_2-\underset{X}{CH}-CH_2$$

termination stage the final step in a polymerization (↑) chain reaction (p.206) in which the active polymer chain ceases to propagate as a result of either two chains combining together, or by ending the addition processes by the supply of further free radicals to the polymer mixture. The length of the polymer (↑) chain can be predetermined by the number of free radical producing molecules added to the reaction. **terminate** (*v*).

termination stage

$$R-(\underset{X}{CH}-CH_2)_n-\underset{X}{CH}-CH_2^\bullet + R^\bullet \longrightarrow R-(\underset{X}{CH}-CH_2)_n-\underset{X}{CH}-CH_2-R$$

addition polymerization the type of polymerization (↑) occurring when bonds in unsaturated compounds, such as alkenes (p.34), are broken to give radicals which can combine through the propagation stage (↑) of the polymerization process. Polypropylene, polychloroprene and polystyrene are all examples of

addition polymerization

$$\mathbf{n} \ \underset{CH_3}{CH}=CH_2 \longrightarrow \left(\underset{CH_3}{CH}=CH_2\right)_\mathbf{n}$$
propylene (propene)   polypropylene

condensation polymerization refers to polymerization processes in which monomers (p.236) are joined together following condensation reactions (p.221) in which a simple molecule, such as water or ethanol, is eliminated. Most condensation polymerizations involve two different molecular species and represent a type of copolymerization (↓), e.g. nylon 66 is a condensation polymer from adipic acid and hexamethylenediamine.

$$HOOC\,(CH_2)_4\,COOH + H_2N\,(CH_2)_4\,NH_2$$

$$\sim\!\!\!\sim\!\!C-(CH_2)_4-C-NH(CH_2)_6\,NH\sim\!\!\!\sim$$
$$\qquad\;\;\overset{\|}{O}\qquad\qquad\;\overset{\|}{O}$$

condensation polymerization

copolymerization (*n*) any polymerization (p.236) process in which two or more monomers (p.236) are reacted together. Most condensation polymerizations (↑) are of this type and involve a ratio of 1:1 for the two monomers. Copolymerizations involving addition processes can more easily use mixed ratios of monomers, e.g. butadiene and styrene can be copolymerized in the ratio 3:1 to produce the synthetic rubber SBR.

$$CH_2 = CH - CH = CH_2 \;+\; CH = CH_2$$

1,3-butadiene   styrene

$$- CH_2 = CH = CH - CH_2 - CH - CH_2 -$$

copolymerization

ionic polymerization refers to chain reaction (p.206) polymerizations (p.236) in which the initiation stage (p.236) and propagation stages (p.237) proceed by the action of ions rather than free radicals (p.214). Initiation (p.236) may be by either cations or anions.

homogeneous ionic polymerization describes ionic polymerization (↑) reactions carried out in a single phase (p.159) system, usually reactions in solution.

linear condensation polymer a condensation polymer formed from monomers which only possess two functional groups such that chain growth can only occur in the two directions at either end of the chain.

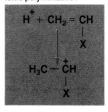

ionic polymerization

$$H^+ + CH_2 = \overset{\displaystyle CH}{\underset{\displaystyle X}{|}}$$

$$H_3C - \overset{+}{\underset{\displaystyle X}{\overset{\displaystyle CH}{|}}}$$

$$2\left(H - \overset{\overset{\displaystyle H}{|}}{C} = O\right) + H_2NCNH_2$$
$$\underset{\overset{\|}{O}}{}$$

$$\downarrow$$

$$HO - CH_2 - NH - \underset{\overset{\|}{O}}{C} - NH - CH_2 - OH$$

$$\xrightarrow[\overset{H_2NCNH_2}{\underset{\overset{\|}{O}}{}}]{H_2C=O}$$

$$\begin{array}{l} \qquad\qquad\qquad\qquad\quad CH_2 \\ \qquad\qquad\qquad\qquad\quad | \\ CH_2 - NH - \underset{\overset{\|}{O}}{C} - N - CH_2 - O - \overset{|}{\underset{|}{C}} - \\ | \\ NH \\ | \\ \underset{\overset{\|}{O}}{C} - NH - CH_2 - O - CH_2 - N - CH_2 \\ \quad\; | \qquad\qquad\qquad\qquad\qquad | \\ \quad CH_2 \quad\text{network polymer}\quad C = O \\ \quad\; | \qquad\qquad\qquad\qquad\qquad | \end{array}$$

network polymer a highly cross-linked polymer such as can be formed by condensation polymerization (p.236) with a polyfunctional chemical, e.g. urea and formaldehyde form a network polymer as each $-NH_2$ group on the urea molecules can condense with two formaldehyde molecules.

link (n) either a bond or an interconnecting atom or group between two polymer chains (p.33).

cross-link (v) to form a bond or establish an interconnecting group between two or more polymer chains. Cross-linking leads to an increase in durability and density of polymers, e.g. sulphur is used in vulcanizing (p.240) rubber, the sulphur forming bridges by reacting with carbon atoms in the natural rubber chains.

binding (n) a term given to the ability of polymer chains to hold together and form stable structures.

natural polymers refers to polymers (p.236) which may be obtained from natural sources, e.g. cellulose, rubber, starch, proteins.

curing (n) the process of making natural rubber harder by vulcanization with sulphur forming cross-links (↑) between polymer chains (p.33). **cure** (v).

gel time the period required to form a soft gel as the initial stage in the strengthening of polymers by the use of glass fibre or reinforced plastics (p.240).

number average degree of polymerization as the length of the chains (p.33) obtained from a polymerization reaction varies, often very greatly, the number average refers to the average number of monomer (p.236) units in each chain. This is given the symbol M_n. The former symbol is DP_n.

weight average degree of polymerization similar to the number average (↑); it refers to the average relative molecular mass (p.29) for the polymer (p.236) chains. It is given the symbol M_w. The former symbol is DP_w.

amorphous polymer a polymer totally lacking in crystal-linity, with a random arrangement of the polymer chain.

plastic (*n*) refers to polymers (p.236) which may either be shaped after being softened by heating, or can be moulded by heat and pressure. *See* **thermoplastic** (↓) and **thermosetting plastic** (↓).

plasticity (*n*) the property of being plastic (↑) at some stage in the process of manufacture of a polymer (p.236), or of an object made from a polymer.

plasticizer (*n*) a chemical which when added to a polymer enables it to retain its plasticity (↑), or makes it plastic (↑).

thermoplastic (*n*) a polymer which becomes soft and can be moulded when it has been heated. Thermoplastics can be re-shaped if heated and moulded again. **thermoplasticity** (*adj*).

thermosetting plastic a polymer, or mixture of substances which will produce a polymer, which is initially plastic (↑) when heated and can be shaped and moulded, but once it has cooled down it cannot be remoulded as it loses its plasticity (↑).

rubbers (*n*) natural and synthetic polymers (p.236) possessing a high degree of unsaturation due to double bonds in the structures which impart elasticity to the molecules. Natural rubber is cis-1,4-polyisoprene, the properties of which are improved by vulcanization (↓). Synthetic rubbers are copolymers of butadiene with either isoprene or styrene.

vulcanization (*n*) the process of forming cross-linkages between the molecular chains in rubber (↑) by heating with sulphur to make the rubber harder and more durable.

glass transition temperatures refers to a short range of temperatures at which molecules in a polymer (p.236) become free to move and the polymer passes from a hard to a soft state. The symbol is T_g.

melt temperature refers to the temperature at which the attractive forces between polymer chains are weakened sufficiently by heat for the softened material to flow. The symbol is T_m.

glasses (*n*) transparent or translucent materials which are non-crystalline and permit the transmission of light. Common glass is made by fusing silica, SiO_2, with sodium carbonate and calcium carbonate; for special purposes salts of other metals, e.g. lead, boron or aluminium may be added. Some polymers (p.236) are also glasses, e.g. poly(methylmethacrylate).

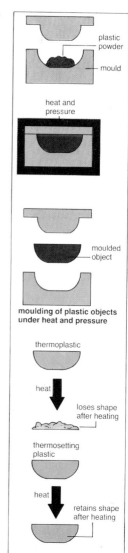

plastic powder

mould

heat and pressure

moulded object

moulding of plastic objects under heat and pressure

thermoplastic

heat

loses shape after heating

thermosetting plastic

heat

retains shape after heating

Physical quantities and S.I. units

Basic units

physical quantity	unit	symbol
length	metre	m
mass	kilogram	kg
time	second	s
electric current	ampere	A
temperature	kelvin	K
amount of substance	mole	mol
luminous intensity	candela	cd

Some prefixes commonly used in S.I. measurements

prefix	fractions	symbol
deci	$\times 10^{-1} = 0.1$	d
centi	$\times 10^{-2} = 0.01$	c
milli	$\times 10^{-3} = 0.001$	m
micro	$\times 10^{-6} = 0.000\,001$	μ
nano	$\times 10^{-9} = 0.000\,000\,001$	n
pico	$\times 10^{-12} = 0.000\,000\,000\,001$	p

prefix	multiples	symbol
tera	$\times 10^{12} = 1\,000\,000\,000\,000$	T
giga	$\times 10^{9} = 1\,000\,000\,000$	G
mega	$\times 10^{6} = 1\,000\,000$	M
kilo	$\times 10^{3} = 1\,000$	k
hecto	$\times 10^{2} = 100$	h
deca	$\times 10^{1} = 10$	da

Physical quantities and S.I. units

quantity	symbol	S.I. unit	symbol
atomic number	A	(a number)	—
charge, electric	Q	coulomb	C
charge on electron	e	coulomb	C
concentration	c	$mol\,dm^{-3}$	—
conductivity	K	$ohm^{-1}\,metre^{-1}$	$\Omega^{-1}m^{-1}$
conductivity, molar	Λ	$ohm^{-1}\,metre^{-1}$	$\Omega^{-1}m^{-1}$
current, electric	I	ampere	A
decay constant	λ	s^{-1}	—
degree of dissociation	α	(a ratio)	—
degree of hydrolysis	h	(a ratio)	—
degree of ionization	α	(a ratio)	—
dielectric constant	E	(a ratio)	—
efficiency	η	(a ratio)	—
electrochemical equivalent	Z	gC^{-1}	—
electromotive force	E	volt	V
energy, internal	U	joule	J
energy, kinetic	E_k	joule	J
energy, potential	E_p	joule	J
enthalpy	H	joule	J
entropy	S	JK^{-1}	—
free energy, Gibbs	G	joule	J
free energy, Helmholz	A	joule	J
frequency	ν	hertz	Hz
half-life	$t_{1/2}$	second	s
heat capacity	C	JK^{-1}	—
heat capacity, molar	c	$JK^{-1}\,mol^{-1}$	—
heat, quantity of	q	joule	J
heat of reaction	ΔH	joule	J
intensity (of e.m. waves)	I	watt	W
latent heat, molar	l_m	joule	J
mass number	A	(a number)	—
molar volume	V_m	dm^3	—
mole fraction	n	(a number)	—
neutron number	N	(a number)	—
number of molecules	N	(a number)	—
permittivity, relative	ε_r	(a ratio)	—
potential, electric	V	volt	V
potential difference	V	volt	—
potential, chemical	μ	$Jmol^{-1}$	—
pressure	p	pascal (Nm^{-2})	Pa
resistivity, electrical	ρ	ohm-metre	Δm
surface tension	γ	Nm^{-1}	—
temperature, absolute	T	kelvin	K
temperature, Celsius	θ	degree Celsius	°C
temperature interval	θ	degree Celsius or kelvin	°C or K
time	t	second	s
velocity	$u.v$	ms^{-1}	—
velocity, e.m. waves	c	ms^{-1}	—
vapour density	Δ	(a ratio)	—
viscosity, coefficient of	η	$kgm^{-1}s^{-1}$	—
wavelength	λ	metre	m
work	w	joule	J

Some useful abbreviations and constants

Common abbreviations

abs.	absolute	insol.	insoluble
a.c.	alternating current	i.r.	infrared
anhyd.	anhydrous	liq.	liquid
a.p.	atmospheric pressure	m.p.	melting point
aq.	aqueous	p.d.	potential difference
b.p.	boiling point	ppt.	precipitate
conc.	concentrated	r.a.m.	relative atomic mass
concn.	concentration	r.d.	relative density
const.	constant	sol.	soluble
crit.	critical	soln.	solution
cryst.	crystalline	s.t.p.	standard temperature and pressure
d.c.	direct current		
dil.	dilute	temp.	temperature
dist.	distilled	u.v.	ultraviolet
e.m.f.	electromotive force	v.d.	vapour density
f.p.	freezing point	V.R.	velocity ratio
h.	hour	w.	weight
hyd.	hydrated		

Physical constants

PHYSICAL CONSTANT	SYMBOL	VALUE
Avogadro constant	No	$6.02 \times 10^{23} \, mol^{-1}$
atomic mass unit	a.m.u.	$1.660 \times 10^{-27} \, kg$
proton mass	1.007 a.m.u.	$1.673 \times 10^{-27} \, kg$
neutron mass	1.009 a.m.u.	$1.675 \times 10^{-27} \, kg$
gas constant		$8.314 \, JK^{-1} mol^{-1}$
electron-volt	eV	$1.60 \times 10^{-19} \, J$
electronic charge	e	$1.60 \times 10^{-19} \, C$
Faraday constant	F	$9.65 \times 10^4 \, Cmol^{-1}$
Planck's constant	h	$6.62 \times 10^{-34} \, Js$
one calorie	C	$4.18 \, J$
electron mass	m	$9.11 \times 10^{-31} \, kg$
s.t.p.		1.00 atm or 760 mmHg or 101 KPa or 0°C or 273.15 K
temperature triple point water		273.16 K
standard volume of mole of gas at s.t.p.		$22.4 \, dm^3$
specific heat capacity water		$4.18 \, Jg^{-1} K^{-1}$

The Periodic Table

The periodic table classifies the elements according to their atomic number. The elements are arranged in groups (arranged vertically) or periods (arranged horizontally). There is a scientific connection between elements within one group or period, e.g. the elements of fluorine, chlorine and bromine are members of the group of elements called halogens; they are some of the most highly reactive elements.

KEY

- HYDROGEN
- ALKALI AND ALKALINE EARTH METALS
- METALS
- NON METALS INCLUDING HALOGENS
- NOBLE GASES

1	2	3	4	5	6	7	8	9	10	11	12	13	14	15	16	17	18
H —																	He 4
Li 7	Be 9											B 11	C 12	N 14	O 16	F 19	Ne 20
Na 23	Mg 24											Al 27	Si 28	P 31	S 32	Cl 35.5	Ar 40
K 39	Ca 40	Sc 45	Ti 48	V 51	Cr 52	Mn 55	Fe 56	Co 59	Ni 59	Cu 64	Zn 65	Ga 70	Ge 73	As 75	Se 79	Br 80	Kr 84
Rb 85	Sr 88	Y 89	Zr 91	Nb 93	Mo 96	Tc 98	Ru 101	Rh 103	Pd 106	Ag 108	Cd 112	In 115	Sn 119	Sb 122	Te 128	I 127	Xe 131
Cs 133	Ba 137	La 139	Hf 178	Ta 181	W 184	Re 186	Os 190	Ir 192	Pt 195	Au 197	Hg 201	Tl 204	Pb 207	Bi 209	Po 209	At 210	Rn 222
Fr 223	Ra 226	Ac 227															

Element names:
HYDROGEN (H), HELIUM (He), LITHIUM (Li), BERYLLIUM (Be), BORON (B), CARBON (C), NITROGEN (N), OXYGEN (O), FLUORINE (F), NEON (Ne), SODIUM (Na), MAGNESIUM (Mg), ALUMINIUM (Al), SILICON (Si), PHOSPHORUS (P), SULPHUR (S), CHLORINE (Cl), ARGON (Ar), POTASSIUM (K), CALCIUM (Ca), SCANDIUM (Sc), TITANIUM (Ti), VANADIUM (V), CHROMIUM (Cr), MANGANESE (Mn), IRON (Fe), COBALT (Co), NICKEL (Ni), COPPER (Cu), ZINC (Zn), GALLIUM (Ga), GERMANIUM (Ge), ARSENIC (As), SELENIUM (Se), BROMINE (Br), KRYPTON (Kr), RUBIDIUM (Rb), STRONTIUM (Sr), YTTRIUM (Y), ZIRCONIUM (Zr), NIOBIUM (Nb), MOLYBDENUM (Mo), TECHNETIUM (Tc), RUTHENIUM (Ru), RHODIUM (Rh), PALLADIUM (Pd), SILVER (Ag), CADMIUM (Cd), INDIUM (In), TIN (Sn), ANTIMONY (Sb), TELLURIUM (Te), IODINE (I), XENON (Xe), CAESIUM (Cs), BARIUM (Ba), LANTHANUM (La), HAFNIUM (Hf), TANTALUM (Ta), TUNGSTEN (W), RHENIUM (Re), OSMIUM (Os), IRIDIUM (Ir), PLATINUM (Pt), GOLD (Au), MERCURY (Hg), THALLIUM (Tl), LEAD (Pb), BISMUTH (Bi), POLONIUM (Po), ASTATINE (At), RADON (Rn), FRANCIUM (Fr), RADIUM (Ra), ACTINIUM (Ac)

The meaning of words in chemistry

Words in chemistry, and science generally, are made from small groups of letters, with each group having a meaning. These groups of letters can be placed at the beginning of a word, or at the end, or they can form the middle, or base, of a word. A group of letters at the front of a word is a **prefix**, and at the end is a **suffix**. The base of a word, in many cases, cannot be used by itself. Words made up from parts can be split into the original parts and the meaning of the whole word made clearer.

Example (a) (1) exist → *existence* (base plus suffix)
(2) exist → *coexist* (base plus prefix)
(3) exist → *coexistence* (prefix and suffix added)

(b) **polymerization** can be split into:

poly-	prefix meaning 'many'
mer–	word base meaning 'a part'
–ize	suffix forming a verb from **polymer**
–ation	suffix meaning 'a process'
polymer	something made from many parts
polymerize	to make something from many parts
polymerization	the process of making something from many parts

PREFIX/ SUFFIX/ BASE	MEANING	EXAMPLE	MEANING
a-/an-	without	aprotic	without a proton
-ance	the property of an object	conductance	the measured ability of a particular electrolyte or conductor to pass current
anti-	opposite in place or effect	anti-bonding	the opposite of bonding
aqua-	water	aqueous	possessing water
auto-	self	auto-ionization	ionizing itself as water does
chrom-	colour	monochromatic	describes something of one colour only
co-	sharing, or being together	coexistence	existing together
de	an opposite action, to remove	dehydration	the process of removing water
dent	tooth, projection	monodentate	having one point of action as a characteristic
dia	through or across	dialysis	breaking up by pressing through by crystalloid pressing through a membrane
endo	in, inside	endothermic	taking heat in
-er	an agent	plasticizer	an agent that makes a substance plastic
exo-	out, outwards	exothermic	giving out heat

PREFIX/ SUFFIX/ BASE	MEANING	EXAMPLE	MEANING
-gram	a result, written or drawn	polarogram	a record of change drawn on paper from the action of the poles of a cell
-graph	an instrument for recording results	barograph	a barometer that records, in drawing, atmospheric pressure
halo	salty	halogen	salt producers, elements similar to chlorine
hygro	wet, humid	hygrometry	the measurement of humidity
-ic	has the property of	ionic	has the properties of ion
in-	not (used with adjectives) [changes to im- before m/b/p, to il- before l, to ir- before o]	insoluble	not soluble
-ivity/-ity	a quality, depending on conditions, of materials	conductivity	the quality of conducting electric current if an e.m.f. is applied to a material
leuco-	white	leuco base	a white compound derived from a coloured compound
meta-	after, behind	metamorphic	after having a change of structure
-meter	an accurate measuring instrument	ammeter	an instrument that measures electric current accurately in amperes
non-	not, opposite in nature (used with nouns and adjectives)	non-ideal	not ideal, with different properties from ideal
pseudo	false, appears what it is not	pseudo-first	a false first, although it appears as first
-scope	instrument used for observation but not measurement	spinthariscope	an instrument for approximate observation of radiation
sub-	beneath, smaller in importance or degree	sub-atomic	smaller than atomic, i.e. describes particles in atoms
super-	greater in quantity or quality	super-cooled	cooled below the normal temperature for liquefaction or solidification
un-	not (used with adjectives)	uncombined	not combined

Index